Springer Theses

Recognizing Outstanding Ph.D. Research

For further volumes:
http://www.springer.com/series/8790

Aims and Scope

The series "Springer Theses" brings together a selection of the very best Ph.D. theses from around the world and across the physical sciences. Nominated and endorsed by two recognized specialists, each published volume has been selected for its scientific excellence and the high impact of its contents for the pertinent field of research. For greater accessibility to non-specialists, the published versions include an extended introduction, as well as a foreword by the student's supervisor explaining the special relevance of the work for the field. As a whole, the series will provide a valuable resource both for newcomers to the research fields described, and for other scientists seeking detailed background information on special questions. Finally, it provides an accredited documentation of the valuable contributions made by today's younger generation of scientists.

Theses are accepted into the series by invited nomination only and must fulfill all of the following criteria

- They must be written in good English.
- The topic should fall within the confines of Chemistry, Physics, Earth Sciences, Engineering and related interdisciplinary fields such as Materials, Nanoscience, Chemical Engineering, Complex Systems and Biophysics.
- The work reported in the thesis must represent a significant scientific advance.
- If the thesis includes previously published material, permission to reproduce this must be gained from the respective copyright holder.
- They must have been examined and passed during the 12 months prior to nomination.
- Each thesis should include a foreword by the supervisor outlining the significance of its content.
- The theses should have a clearly defined structure including an introduction accessible to scientists not expert in that particular field.

Jonathan C. F. Matthews

Multi-Photon Quantum Information Science and Technology in Integrated Optics

Doctoral Thesis accepted by
the University of Bristol, UK

 Springer

Author
Dr. Jonathan C. F. Matthews
H. H. Wills Physics Laboratory
Department of Electrical and
  Electronic Engineering
Centre for Quantum Photonics
University of Bristol
Bristol
UK

Supervisor
Prof. Jeremy L. O'Brien
H. H. Wills Physics Laboratory
Department of Electrical and
  Electronic Engineering
Centre for Quantum Photonics
University of Bristol
Bristol
UK

ISSN 2190-5053
ISBN 978-3-642-44498-2
DOI 10.1007/978-3-642-32870-1
Springer Heidelberg New York Dordrecht London

ISSN 2190-5061   (electronic)
ISBN 978-3-642-32870-1   (eBook)

Publications

Papers

The following papers report work that took place between the beginning of my Ph.D. studies and completion of this thesis. Some of the results reported in this thesis and abbreviated discussion thereof are published in references marked with the corresponding chapter.

1. {Chapter 5} J. C. F. Matthews, A. Politi, A. Stefanov, J. L. O'Brien, "Manipulation of multiphoton entanglement in waveguide quantum circuits", *Nature Photonics*, **3**, 346–350 (2009).
2. {Chapter 3} G. D. Marshall, A. Politi, J. C. F. Matthews, P. Dekker, M. Ams, M. J. Withford, J. L. O'Brien, "Laser written waveguide photonic quantum circuits", *Optics Express*, **17**, 15, 12546–12554, (2009).
3. {Chapter 4} A. Politi, J. C. F. Matthews, J. L. O'Brien, Shor's "Quantum Factoring Algorithm on a Photonic Chip", *Science*, **325**, 1221 (2009).
4. {Chapter 2} A. Politi, J. C. F. Matthews, M. G. Thompson, J. L. O'Brien, "Integrated Quantum Photonics", *IEEE Journal of Selected Topics in Quantum Electronics*, **15**, 6, 1673–1684, (2009).
5. {Chapter 7} A. Peruzzo, M. Lobino, J. C. F. Matthews, N. Matsuda, A. Politi, K. Poulios, X.-Q. Zhou, Y. Lahini, N. Ismail, K. Worhoff, Y. Bromberg, Y. Silberberg, M. G. Thompson, J. L. O'Brien, "Quantum Walks of Correlated Photons", *Science* **329**, 1500–1503 (2010).
6. M. G. Thompson, A. Politi, J. C. F. Matthews, J. L. O'Brien, "Integrated waveguide circuits for optical quantum computing", *IET Circuits Devices Syst.*, **5**, 2, 94–102, (2011).
7. G. D. Marshall, T. Gaebel, J. C. F. Matthews, J. Enderlein, J. L. O'Brien, J. R. Rabeau, "Coherence properties of a single dipole emitter in diamond", *New J. Phys.* **13**, 055016 (2011).
8. {Chapter 8} J. C. F. Matthews, K. Poulios, J. D. A. Meinecke, A. Poilit, A. Peruzzo, N. Ismail, K. Wörhoff, M. G. Thompson, J. L. O'Brien,

"Simulating quantum statistics with entangled photons: a continuous transition from bosons to fermions", *arXiv:1106.1166 [quant-ph], submitted.*

9. {Chapter 6} J. C. F. Matthews, A. Politi, D. Bonneau, J. L. O'Brien, "Heralding Two-Photon and Four-Photon Path Entanglement on a Chip", *Phys. Rev. Lett.* **107**, 163602 (2011).
10. P. J. Shadbolt, M. R. Verde, A. Peruzzo, A. Politi, A. Laing, M. Lobino, J. C. F. Matthews, M. G. Thompson, J. L. O'Brien, "Generating, manipulating and measuring entanglement and mixture using a reconfigurable integrated quantum photonic circuit", *Nature Photonics* **6**, 45–49 (2012).
11. A. Crespi, M. Lobino, J. C. F. Matthews, A. Politi, C. R. Neal, R. Ramponi, R. Osellame, J. L. O'Brien, "Measuring protein concentration with entangled photons", *Applied Physics Letters* **100**, 233704 (2012).

Conference Talks Presented

In addition to the following personal presentations, I have been a co-author of more than 60 talks at international conferences including invited and post-deadline talks.

1. J. C. F. Matthews, K. Poulios, J. Meinecke, A. Politi, N. Ismail, K. Worhoff, M. G. Thompson, J. L. O'Brien, Bristol Physics Graduate School Colloquium, 24/01/11.
2. J. C. F. Matthews, A. Politi, D. Bonneau, A. Stefanov, J. L. O'Brien, "Entangled multi-photon states in waveguide for quantum metrology", Physics of Quantum Electronics (Quantum sensing session), Snowbird, Utah, 05/01/11 (Invited).
3. J. C. F. Matthews, A. Politi, D. Bonneau, A. Stefanov, J. L. O'Brien, "Entangled multi-photon states in waveguide for quantum metrology", Photon 10/QEP-19 (joint session), Southampton, 25/08/10 (Contributed).
4. J. C. F. Matthews, A. Politi, A. Peruzzo, A. Laing, P. Kalasuwan, X.-Q. Zhou, M. Rodas Verde, D. Bonneau, A. Stefanov, T. C. Ralph, N. Matsuda, N. Ismail, K. Worhoff, Y. Lahini, Y. Bromberg, Y. Silberberg, J. G. Rarity, J. L. O'Brien, "Quantum Information Science With Photonic Chips", CEWQO 2010, St Andrews, 6 June 2010 (Contributed).
5. J. C. F. Matthews, A. Politi, D. Bonneau, A. Stefanov, J. L. O'Brien, "Entangled multi-photon states in waveguide for quantum metrology", ECIO 2010, Cambridge, 9 April 2010 (Contributed).
6. J. C. F. Matthews, A. Politi, J. L. O'Brien, "A Compiled Version of Shor's Quantum Factoring Algorithm on a Waveguide Chip", Frontiers in Optics 2009, San Jose, 14 Oct 2009 (Post-deadline session).
7. J. C. F. Matthews, A. Peruzzo, A. Politi, A. Laing, P. Kalasuwan, X.-Q. Zhou, M. Rodas Verde, M. J. Cryan, J. G. Rarity, A. Stefanov, S. Yu, M. G. Thompson, J. L. O'Brien, "Quantum Information Science with Photons on a Chip", Frontiers in Optics 2009, San Jose, 12 Oct 2009 (Invited).

8. J. C. F. Matthews, A. Politi, A. Stefanov, J. L. O'Brien, "Manipulating Photonic Quntum States and a Compiled Quantum Algorithm Using Silica Waveguides", QIPIRC Quanutm Photonics Workshop, 18 Sept 2009 (Contributed).

9. J. C. F. Matthews, A. Politi, A. Stefanov, J. L. O'Brien, "Integrated Quantum Information Science with Photons", Bristol Physics Department, Postgraduate conference, 19 June 2009 (Contributed).

10. J. C. F. Matthews, A. Politi, A. Laing, A. Peruzzo, P. Kalasuwan, M. Zhang, X.-Q. Zhou, M. Rodas Verde, A. Stefanov, M. J. Cryan, J. G. Rarity, S. Yu, M. H. Thompson, J. L. O'Brien, "Integrated Quantum Information Science with Photons", CLEO, Baltimore, MD, USA, June 2009 (Post-deadline session).

11. J. C. F. Matthews, A. Politi, A. Laing, A. S. Clark, J. Fulconis, A. Stefanov, M. J. Cryan, J. G. Rarity, T. Ruddolph, S. Yu, G. D. Marshall, M. Ams, P. Dekker, M. J. Withford, J. L. O'Brien, "Fabricating and Identifying Photonic Quantum Devices", ICQO, Vilnius, Lithuania, September 2008 (invited).

12. J. C. F. Matthews, A. Politi, A. Laing, A. S. Clark, J. Fulconis, A. Stefanov, M. J. Cryan, J. G. Rarity, T. Rudolph, S. Yu, G. D. Marshall, M. Ams, P. Dekker, M. Withford, J. L. O'Brien, "Quantum Photonics on a Chip", SPIE Optics & Photonics, San Diego, CA, USA, August 2008 (invited).

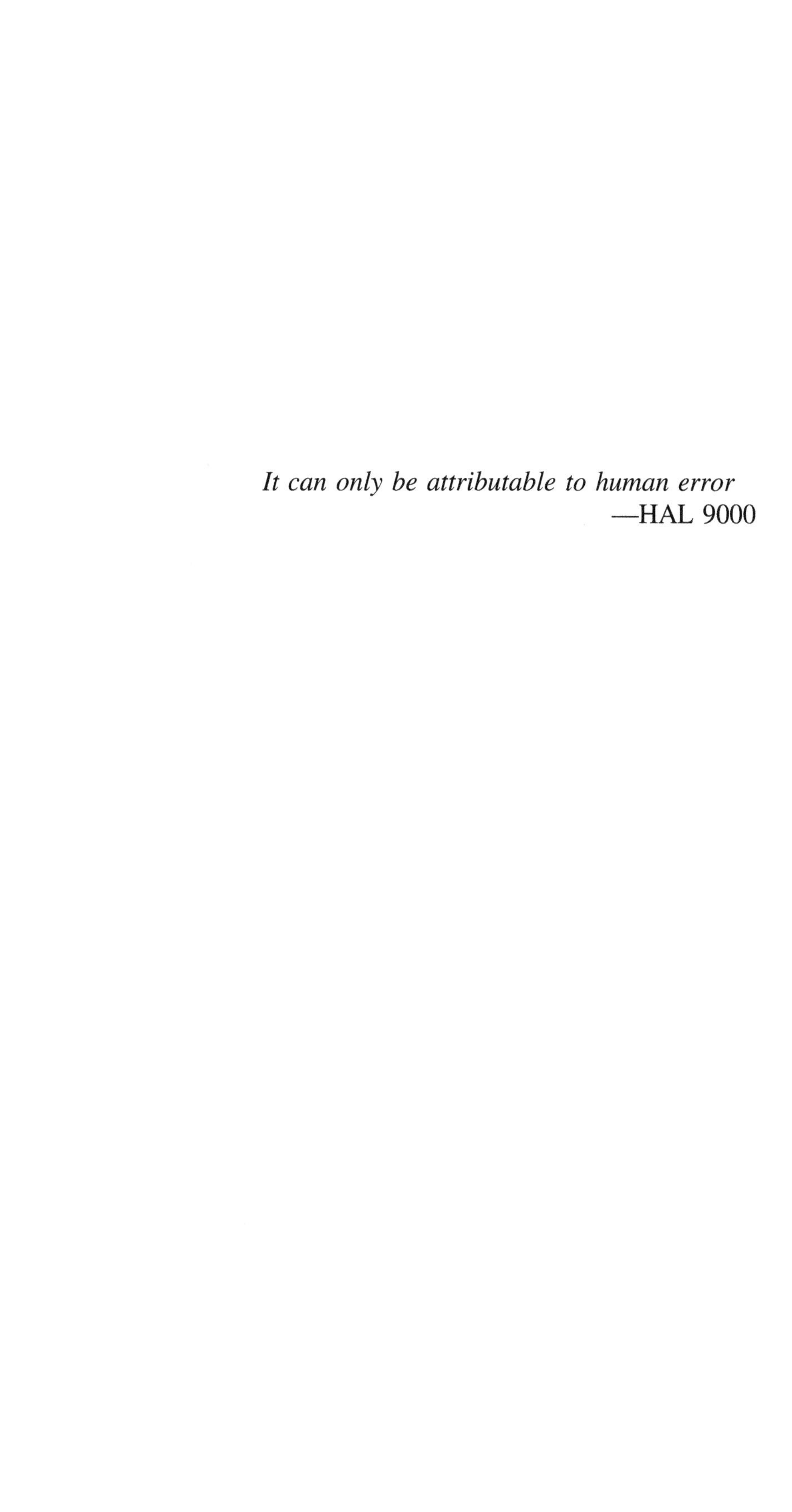

It can only be attributable to human error
—HAL 9000

Supervisor's Foreword

This thesis reports pioneering results in *Integrated Quantum Photonics*, a field that is revolutionising quantum science, by harnessing the miniature scale and inherent stability offered by integrated optics to realise scalable quantum circuitry with unprecedented complexity. Manipulating the quantum state of multiple photons (particles of light) is an ideal platform with which to observe and test quantum mechanics and is a promising approach to develop disruptive technology that is fundamentally enhanced by unique properties of single photons, including computing, simulation and precision measurement. However, photonic circuits developed for these tasks were rapidly approaching limits in stability and complexity using traditional optical elements bolted onto optical tables. The results presented are foundational for a new paradigm in quantum optics that realises photonic circuits in monolithic chip architectures. Key building blocks for fully integrated quantum optical technology are demonstrated that enable future photonic quantum optical technology and open the way to novel avenues of research and development. Highlights include rapidly prototyped laser-written quantum circuits; the first quantum algorithm realised on a chip; reconfigurable quantum circuitry; the first multi-particle quantum walk and physical simulation of fermion behaviour using entangled photons. These represent important advances in contemporary quantum physics and have already enabled developments in highly reconfigurable quantum processors (e.g. [Nature Photon. 6, 45–49 (2012)]), and practical quantum-enhanced sensors (e.g. [Appl. Phys. Lett. 100, 233704 (2012)]).

Bristol, 2012 Jeremy L. O'Brien

Acknowledgments

I am grateful to a lot of people for making my Ph.D. studies such an exciting and happy time. I have been privileged to work in a stimulating environment with a group of friendly and bright people. I thank my supervisor Jeremy O'Brien and John Rarity for establishing the group and thank you both for your faith in me and for giving me the chance to make my contribution. I thank Ruth Oulton and Mark Thompson for their continued efforts in running the group alongside John and Jeremy.

Jeremy has taught me more than I would have ever expected or hoped from a Ph.D. His enthusiasm, his instruction, his trust and his encouragement has been instrumental in my development as a physicist and as a writer. He has provided insight and superb instruction on leading scientific research. He has encouraged me to travel to international conferences and research groups, allowing me to meet a vast number of like-minded individuals and field leaders in the community. Thank you for the thought you have put into your group, your understanding, for the numerous opportunities, the good humour and for all the beer.

I have had the pleasure of making a good friend in Alberto Politi. Alberto is a smart and knowledgeable man. We have had countless terrific discussions that have mostly led to great work and always to important understanding. He has taught me an awful lot, from coupling fibres to waveguide chips, to making proper coffee. I have thoroughly enjoyed our work together. I also note that there are few that know more about *The Simpsons* than Alberto.

Anthony Laing and André Stefanov taught me to walk; they taught how to align optics in the early days and how to get started on what at the time was a rather daunting task of building a multi-photon source. Alistair Lynch and Mark Godfrey get special mention for building great FPGA counting logic, which I have used throughout my Ph.D. and beyond for counting photons.

As I have witnessed the group grow to a size of $\sim$40 members, I have worked in projects along side some exceptionally bright guys: Jasmin Meinecke, Dan Fry, Jake Kennard, Kostas Poulios, Pete Shadbolt, Alberto Peruzzo and Damien Bonneau; and of course two wise postdocs: Xiao-Qi Zhou and Mirko Lobino. I would like to mention Andrew Young and Mark Godfrey's constant banter (and occasional sage advice) that has added to the good sense of humour of the whole group that I will always associate to Bristol, and I would like to mention the grand discussions of life and physics that I have had with many of the group, most notably with J. P. Hadden. Alberto Politi, Jasmin Meinecke and Mark Godfrey gain special thanks for proof reading my thesis for typographical errors.

I thank my skiing instructors and teammates in the photonics ski team over the years, in particular JGR for his organisation of a great annual trip. My Physics skittles teammates—"Brownian Motion"—have my thanks for entertaining and competitive evenings in the early years and it was an honour to captain for a season and keeping this great sport alive; "let it roll".

From joining our group in the summer of 2011, Rebecca Morton has already organised and revolutionised the way we operate. She has been particularly great in organising my thesis submission and my examination. A lovely personality, a great addition to the group. It is hard to imagine how we operated without her.

To all the Italians in the group: nice coffee, in particular Luca Marseglia. Thank you for keeping me from snoozing in the afternoon. Special thanks go to Helen and Krishna Nathan for looking after us in the Green Man with great beer and lovely food, certainly one of my favourite places to talk physics.

I would like to thank James Rabeau, Graham Marshall, Jason Twamley and their colleagues at Macquarie University, for making feel so welcome during my 2 month visit to Australia in 2009, yielding some great collaborative work.

Finally I would like to thank my family. I will always look up to my big brother Peter, he is the chap responsible for sparking my interests and opening my mind to the beauty of physics and mathematics, right from an early age. This amplified my pride in showing him our labs and demonstrating Hong–Ou–Mandel interference during his visit to the UK in 2010. Peter, big sister Nicola and my parents Anne and David, have together always encouraged my interests in science (albeit with promises that it would lead to owning a yellow Ferrari when I grew up; this has not happened yet). Now the next generation in my family provide me with new perspectives on science—to my niece, Charlotte, and my nephews Daniel, Frederick and Maxwell, I thank them for their viewpoints on science; I have particularly enjoyed discussing lasers and robots. Finally, I thank my parents for giving me the much needed support throughout my studies and my research; thank you for all the Sunday roasts, home cooked currys and cups of tea during my visits back home. Thank you for the continued support and encouragement.

Bristol, 2011   Jonathan C. F. Matthews

Contents

Chapter 1
Introduction

Fundamental quantum physics and quantum information science has found great experimental success through control over single photons in optical circuits. To date, however, the majority of quantum optical experiments use large scale (bulk) optical elements bolted down to an optical bench; an approach that ultimately limits the complexity and stability of quantum circuits required for quantum science and technology (QST). Here, a series of experiments are reported in the emerging field of integrated quantum photonics that show monolithic waveguide chips to be a suitable platform for realising the next generation of quantum optical circuits. These demonstrations are foundational in developing a new quantum photonic platform necessary for studying fundamental quantum physics and for advancing quantum information science and technology.

1.1 Introduction to Quantum Information Science

Quantum mechanics predicts that nature, at the atomic and sub-atomic level, behaves in a way that is counter-intuitive to our conventional macroscopic experience. A single particle can exist simultaneously in multiple places in *superposition* allowing wave-like interference. However, when the particle is observed, its state 'collapses' to being at only one position with probability, implying any single quantum event in general cannot be predicted with certainty. Perhaps even more startling is the phenomena of entanglement; two or more quantum particles can be separated in space, and yet still share a connection via their combined quantum state, such that measurement on one particle instantaneously effects the quantum state of the other. Such quantum mechanical phenomena has been observed with high precision and has ignited a worldwide effort to harness the unique quantum mechanical phenomena of superposition and entanglement to fundamentally enhance both technology and understanding of the natural world.

J. C. F. Matthews, *Multi-Photon Quantum Information Science and Technology in Integrated Optics*, Springer Theses, DOI: 10.1007/978-3-642-32870-1_1,
© Springer-Verlag Berlin Heidelberg 2013

Under the umbrella of *quantum information science* [1], several quantum technologies have been proposed that are based on encoding and manipulating information using quantum superposition and entanglement. *Quantum cryptography* [2] aims to fundamentally ensure information security using the laws of quantum mechanics and already has commercial companies providing quantum key distribution systems. *Quantum computation* [3] is a field in which algorithms are being designed to run on computers that have access to quantum phenomena and can solve intractable problems such as factorisation—these problems performed otherwise scale so badly on any computer confined to classical physics that they are effectively impossible. In the long term, quantum algorithms will be run on a full scale *universal quantum computer*, that is tolerant to computation and fabrication errors. On such a machine, it will also be possible to accurately simulate complex quantum systems [4] that can only be approximate by any other means. However, in the near term, purpose built quantum machines aimed at an *analogue* form of *quantum simulation* [5] are likely to provide computational advantage over current classical systems. *Quantum metrology* [6] offers the means to fundamentally enhance precision in measurement, beyond the classical shot noise 'limit', by harnessing entanglement to obtain the highest possible precision permitted by nature. With simulation, computation and measurement forming the foundations of science and engineering, the successful development of true quantum technologies presents an exciting prospect.

Multiple challenges must be overcome before fully functional and practical quantum technologies can be realised. These challenges, in particular for computation, equate to the scalable realisation, accurate measurement and control of elements of quantum information known as 'qubits'. *Quantum bits* or *qubits* are the quantum analogue of the classical bit of information and can exist in superposition of the logical "0" and "1" levels as quantum states $|0\rangle$ and $|1\rangle$ respectively; the most general (pure) qubit is given by

$$|\psi\rangle = \cos\theta \, |0\rangle + e^{i\phi} \sin\theta \, |1\rangle \tag{1.1}$$

for angles $0 \leq \theta \leq \pi/2$ and $0 < \phi < \pi$. A specific set of criteria for quantum computation [7] requires qubits to be realised in a physical system such that they can be accurately prepared in an initial state; they must also be individually addressable; qubits must be sufficiency shielded from the environment (such that their quantum nature is preserved long enough for meaningful operation); and it must be possible to perform a set of 'universal quantum logic gates' on qubits that rotate both single- and two-qubit states and can generate entanglement [1].

Many physical quantum platforms are being developed to fulfil the criteria for a viable qubit and for quantum technologies in general, including trapped neutral atoms and trapped ions, solid state, semiconductors, nuclear magnetic resonance, cavity quantum electrodynamics and quantum states of light [1]. One leading approach is the science of single photons—quantum photonics.

1.2 Photonic Quantum Information Science

Photons can almost effortlessly fulfil many of the criteria for encoding quantum information; a single qubit can be encoded on a single photon in many ways, for example the orthogonal horizontal $|H\rangle$ and vertical $|V\rangle$ polarisation states can be used to respectively represent the computational basis $\{|0\rangle, |1\rangle\}$, while arbitrary single qubit states can be prepared and manipulated using commercially available waveplates [8]. Experiments reported in this thesis predominantly uses quantum information encoded in the path of single photons, where for a qubit two separate paths are labeled as '0' and '1' and therefore the presence of a photon again represents the computational basis for a qubit; superposition states can then be prepared using beamsplitters and phase shifts (as we shall see in detail in Chaps. 4 and 5 for example).

Photons also are well known to lack problematic interaction with the environment that would lead to decoherence; this makes photons excellent 'flying qubits' for quantum communication (e.g. [9, 10]). However, photons also do not directly interact with each other leading to the difficulty that optics may not fulfil the criteria of realising two qubit entangling gates. In 2001, however, Knill, Laflamme and Milburn (KLM) [11] proved contrary to common belief that linear optical components, together with single photon sources (including auxiliary photons) and single photon detectors can induce the effective nonlinearity required to create deterministic entangling gates; an all-optical quantum computer is indeed possible according to the laws of quantum mechanics. The resources required for such an implementation are still extremely daunting, however KLM has served as an important inspiration in the field, leading to optical demonstrations of two qubit entangling gates [12, 13], error correction protocols [14–16], compiled versions of Shor's quantum factoring algorithm [17, 18] and proof of principle quantum simulation [19, 20]. Similarly, the potential for photonic implementations of quantum metrology has been illustrated with super-resolved and super-sensitive interferometric experiments using single photons, showing the shot noise limit in precision can be beaten [21–24].

Moreover, photons have proven key in observing, verifying and understanding principal quantum mechanical phenomena on which quantum information science is based, in particular entanglement. Entanglement explains the famous Einstein-Podolsky-Rosen Gedankenexperiment [25] via Bell inequality [26] violations which were first experimentally tested using photons emitted from atomic cascades [27–30] and with entangled photons emitted from spontaneous parametric down conversion [31–33]. These studies have lead onto photonic demonstrations of protocols critically dependent on entanglement, in particular superdense coding (compression of multiple classical bits onto a single qubit) [34], quantum teleportation (central to the KLM scheme) [35, 36] and large entangled systems [37, 38].

1.3 Waveguide Chips for Quantum Optics

Optical implementations of quantum information science and experimental study of quantum mechanics has typically taken place using individual large-scale (bulk) optical elements bolted to optical tables. The use of individual components has certainly been a successful platform and will continue to be a valuable and flexible architecture, allowing precise control and continual development of a given setup. However, as experimental demands on the optical complexity grow, continued activity using only bulk optical architecture will ultimately hinder quantum optical science due to limitations of stability and achievable complexity. Furthermore, implementation of quantum technologies outside of stable laboratory environments requires miniaturisation, a practical architecture and isolation from the surrounding environment.

One promising approach is to miniaturise circuitry using waveguide optics that were primarily designed for the telecommunications industry. Implementation of quantum circuits with optical fibre—as demonstrated with all-fibre polarisation entangling gates [39]—reduces optical bench space required by approximately an order of magnitude compared to bulk optics and, for example, has applications for information processing in quantum optical fibre networks.

Miniaturising circuitry further using using on-chip waveguide circuitry provides a means for stable interferometry and initially has been used for stable time-bin analysis for quantum key distribution at telecommunication wavelengths [40, 41]. The precise mode overlap in waveguide versions of beamsplitters and the inherent stability offered by integration is extremely attractive to achieve the high quality quantum and classical interference required for optical implementations of quantum science and technology.

The first two photon quantum interference experiments in waveguide chips used directional couplers fabricated in silica-on-silicon [42] and a year later in directly laser written waveguides in silica [43] (Chap. 3). The former work [42] also demonstrated single two qubit quantum logic gates circuits, of which two were integrated to perform a proof-of-principle subroutine of Shor's quantum factoring algorithm to factorize 15 [44] (Chap. 4). Near perfect quantum interference and quantum logic gate performance has since been observed in silica-on-silicon waveguides [45]. Control of optical phase of quantum states has been demonstrated in silica-on-silicon [46] (Chap. 5) and in a hybrid approach to fabrication using direct writing with UV lasers [47], using thermal phase shifts aimed at both reconfigurable circuits and manipulation of entangled photonic states suited to quantum metrology. A thermal phase shift with multiple directional couplers has since been used to test a scheme for heralding four photon entanglement from a six photon input [48] (Chap. 6).

All of the above described quantum interference experiments use optical path to encode quantum information (with all other photon properties being made equal). Polarisation encoding has also been explored in directly laser written circuits to expand the use of quantum optical circuits for quantum experiments. Entanglement analysis [49] and most recently, two qubit polarisation entangling gates in waveguide [50] have been demonstrated.

A range of photonic devices exist that do not have a direct analogue to bulk optical elements such as birefringent waveplates and beamsplitters. The mature field of integrated classical photonics continues to be a source of inspiration for integrated quantum photonic devices. Multi-mode interference (MMI) couplers for example, offer a robust means for coupling large numbers of modes together in a single component. Two- and four-input/output MMI couplers have recently been demonstrated for generalised quantum interference [51]. Another photonic component that has found particular application to quantum experiments is evanescently coupled waveguide arrays; these have been shown to be particularly useful for implementing single particle quantum walks using bright light experiments [52, 53]. Such arrays fabricated in planar silicon oxynitride [54] (Chap. 7) and more recently with three-dimensional directly written waveguide structures [55], have been used to realise correlated quantum walks of two photons undergoing quantum interference in the same structure. Furthermore, the birefringent nature of the silicon oxynitride arrays has been exploited to realise two entangled quantum walks and to show that entanglement can exactly simulate arbitrary particle interference in complex networks [56] (Chap. 8).

Observing quantum effects along with demonstrating quantum operation and control directly on monolithic chips will enable new developments in the field of quantum optics and quantum information science in general. As the level of complexity is increased, it is anticipated that advances in all quantum photonic technologies will follow.

1.4 Thesis Summary

During my PhD studies at the University of Bristol, I have performed with collaborators a series of experiments in the realm of integrated quantum photonics, using on-chip waveguide circuitry as the main optical platform. It has not been my role to fabricate nor design waveguide architectures. My main roles have been to assemble and maintain photon sources based on 'parametric downconversion', to use the photon sources to test a range of waveguide circuits with quantum states of light, to analyse data and to design experiments. My thesis is organised into chapters according to the following.

Chapter 2 contains background and experimental details that are common to multiple chapters of the thesis. A model of spontaneous parametric down conversion is reviewed and experimental details are provided for two- and multi-photon down conversion sources used for experiments reported in this thesis. The underlying physical concepts of waveguide optics are discussed and we describe the three waveguide architectures relevant to this thesis—(i) directly laser written waveguides in silica, (ii) doped silica waveguides on a silicon substrate and (iii) silicon oxynitride waveguide arrays.

Chapter 3 reviews the theory of photonic quantum interference and reports the results of two- and three-photon quantum interference using directly written waveguide directional couplers.

Chapter 4 reports a proof-of-principle demonstration of a compiled version (a subroutine) of Shor's quantum factoring algorithm to factorize 15. This is implemented using a waveguide chip of directional couplers to realise single- and two-qubit quantum logic.

Chapter 5 reports quantum interference in an integrated waveguide interferometer with a variable internal phase shift. This component forms the building block for realising large scale, fully reconfigurable quantum networks that would simply be too unstable to realise with bulk optics or optical fibres. We manipulate single photon states on chip, equivalent to single qubit rotations. We observe super-sensitive interference fringes of two- and four-photon number-path entangled 'NOON' states. We also use the device as a reconfigurable beam splitter to realise multiple Hong–Ou–Mandel quantum interference experiments.

In Chap. 6, quantum interference of four- and six-photon states in a waveguide circuit is used to test a scheme for heralding respectively two- and four-photon number-path entangled 'NOON' states, for application to quantum metrology.

Chapter 7 is based on unitary processes known as quantum walks. We discuss the theory of multi-particle quantum walks, in particular their implementation on one-dimensional graphs to simulate the quantum walk of one particle moving around a two-dimensional lattice graph. We report an experimental realisation using two photons undergoing quantum interference in an evanescently coupled waveguide array to realise a two particle quantum walk.

Chapter 8 reports the experimental simulation of arbitrary particle statistics. Using polarisation entanglement undergoing two quantum walks, we observe behaviour associated to bosons, fermions and intermediate fractional behaviour associated to anyons. We show that the experiment generalises for any mode transformation and for an arbitrary number of particles with arbitrary abelian exchange statistics. We provide a proof that the required entanglement can be generated with a circuit that grows polynomially with the number of particles simulated.

Chapter 9 concludes the thesis with a summary of the key results obtained and an outlook to future research.

References

1. M.A. Nielsen, I.L. Chuang, *Quantum Computation and Quantum Information* (Cambridge University Press, Cambridge, 2000)
2. N. Gisin, R. Thew, Quantum communication. Nat. Photonics **1**, 165–171 (2007)
3. T.D. Ladd, F. Jelezko, R. Laflamme, Y. Nakamura, C. Monroe, J.L. O'Brien, Quantum computers. Nature **464**, 45–53 (2010)
4. S. Lloyd, Universal quantum simulators. Science **273**, 1073–1078 (1996)
5. I. Buluta, F. Nori, Quantum simulators. Science **326**, 108 (2009)

6. V. Giovannetti, S. Lloyd, L. Maccone, Quantum-enhanced measurements: beating the standard quantum limit. Science **306**, 1330–1336 (2004)
7. D.P. DiVincenzo, D. Loss, Quantum information is physical. Superlatt. Microstruct. **23**, 419–432 (1998)
8. J.L. O'Brien, Optical quantum computing. Science **318**(5856), 1567–1570 (2007)
9. C. Kurtsiefer, P. Zarda, M. Halder, H. Weinfurter, P.M. Gorman, P.R. Tapster, J.G. Rarity, Quantum cryptography: a step towards global key distribution. Nature **419**(6906), 450–450 (2002)
10. M. Peev, C. Pacher, R. Alleaume, C. Barreiro, J. Bouda, W. Boxleitner, T. Debuisschert, E. Diamanti, M. Dianati, J.F. Dynes, S. Fasel, S. Fossier, M. Fuerst, J.-D. Gautier, O. Gay, N. Gisin, P. Grangier, A. Happe, Y. Hasani, M. Hentschel, H. Huebel, G. Humer, T. Laenger, M. Legre, R. Lieger, J. Lodewyck, T. Loruenser, N. Luetkenhaus, A. Marhold, T. Matyus, O. Maurhart, L. Monat, S. Nauerth, J.-B. Page, A. Poppe, E. Querasser, G. Ribordy, S. Robyr, L. Salvail, A.W. Sharpe, A.J. Shields, D. Stucki, M. Suda, C. Tamas, T. Themel, R.T. Thew, Y. Thoma, A. Treiber, P. Trinkler, R. Tualle-Brouri, F. Vannel, N. Walenta, H. Weier, H. Weinfurter, I. Wimberger, Z.L. Yuan, H. Zbinden, A. Zeilinger, The secoqc quantum key distribution network in vienna. New J. Phys. **11**, 075001 (2009)
11. E. Knill, R. Laflamme, G.J. Milburn, A scheme for efficient quantum computation with linear optics. Nature **409**(6816), 46–52 (2001)
12. J.L. O'Brien, G.J. Pryde, A.G. White, T.C. Ralph, D. Branning, Demonstration of an all-optical quantum controlled-not gate. Nature **426**(6964), 264–267 (2003)
13. S. Gasparoni, J.W. Pan, P. Walther, T. Rudolph, A. Zeilinger, Realization of a photonic controlled-not gate sufficient for quantum computation. Phys. Rev. Lett. **93**, 020504 (2004)
14. T.B. Pittman, B.C. Jacobs, J.D. Franson, Demonstration of quantum error correction using linear optics. Phys. Rev. A **71**, 052332 (2005)
15. J.L. O'Brien, G.J. Pryde, A.G. White, T.C. Ralph, High-fidelity z-measurement error encoding of optical qubits. Phys. Rev. A **71**(6), 060303 (2005)
16. C.Y. Lu, W-B. Gao, J. Zhang, X.-Q. Zhou, T. Yang, J-W. Pan, Experimental quantum coding against qubit loss error. Proc. Natl. Acad. Sci. USA **105**, 11050–11054 (2008)
17. C.-Y. Lu, D.E. Browne, T. Yang, J.W. Pan, Demonstration of a compiled version of shor's quantum factoring algorithm using photonic qubits. Phys. Rev. Lett. **99**, 250504 (2007)
18. B.P. Lanyon, T.J. Weinhold, N.K. Langford, M. Barbieri, D.F.V. James, A. Gilchrist, A.G. White, Experimental demonstration of a compiled version of Shor's algorithm with quantum entanglement. Phys. Rev. Lett. **99**(25), 250505 (2007)
19. B.P. Lanyon, J.D. Whitfield, G.G. Gillett, M.E. Goggin, M.P. Almeida, I. Kassal, J.D. Biamonte, M. Mohseni, B.J. Powell, M. Barbieri, A. Aspuru-Guzik, A.G. White, Towards quantum chemistry on a quantum computer. Nat. Chem. **2**, 106–111 (2010)
20. X.-S. Ma, B. Dakic, W. Naylor, A. Zeilinger, P. Walther, Quantum simulation of the wavefunction to probe frustrated Heisenberg spin systems. Nat. Phys. **7**, 399–405 (2011)
21. M.W. Mitchell, J.S. Lundeen, A.M. Steinberg, Super-resolving phase measurements with a multiphoton entangled state. Nature **429**(6988), 161–164 (2004)
22. P. Walther, J.W. Pan, M. Aspelmeyer, R. Ursin, S. Gasparoni, A. Zeilinger, De broglie wavelength of a non-local four-photon state. Nature **429**(6988), 158–161 (2004)
23. T. Nagata, R. Okamoto, J.L. O'brien, K. Sasaki, S. Takeuchi, Beating the standard quantum limit with four-entangled photons. Science **316**(5825), 726–729 (2007)
24. B.L. Higgins, D.W. Berry, S.D. Bartlett, H.M. Wiseman, G.J. Pryde, Entanglement-free Heisenberg-limited phase estimation. Nature **450**, 393 (2007)
25. A. Einstein, B. Podolsky, N. Rosen, Can quantum-mechanical description of physical reality be considered complete? Phys. Rev. **47**, 777–780 (1935)
26. J. Bell, On the Einstein Podolsky Rosen paradox. Physics **1**, 195–200 (1964)
27. S.J. Freedman, J.F. Clauser, Experimental test of local hidden-variable theories. Phys. Rev. Lett. **28**(14), 938–941 (1972)
28. A. Aspect, P. Grangier, G. Roger, Experimental tests of realistic local theories via bell's theorem. Phys. Rev. Lett. **47**, 460–463 (1981)

29. A. Aspect, P. Grangier, G. Roger, Experimental realization of Einstein-Podolsky-Rosen-Bohm Gedankenexperiment: A new violation of bell's inequalities. Phys. Rev. Lett. **49**(2), 91–94 (1982)
30. A. Aspect, J. Dalibard, G. Roger, Experimental test of bell's inequalities using time-varying analyzers. Phys. Rev. Lett. **49**, 1804–1807 (1982)
31. J.G. Rarity, P.R. Tapster, Experimental violation of bell's inequality based on phase and momentum. Phys. Rev. Lett. **64**, 2495–2498 (1990)
32. G. Whihs, T. Jennewein, C. Simon, H. Weinfurter, A. Zeilinger, Violations of bell's inequality under strict Einstein locality conditions. Phys. Rev. Lett. **81**, 5039–5043 (1998)
33. M.A. Rowe, D. Kielpinski, V. Meyer, C.A. Sackett, W.M. Itano, C. Monroe, D.J. Wineland, Experimental violation of a bell's inequality with efficient detection. Nature **409**, 791–794 (2001)
34. K. Mattle, H. Weinfurter, P.G. Kwiat, A. Zeilinger, Dense coding in experimental quantum communication. Phys. Rev. Lett. **76**, 4656–4659 (1996)
35. D. Bouwmeester, J.-W. Pan, K. Mattle, M. Eibl, H. Weinfurter, A. Zeilinger, Experimental quantum teleportation. Nature **390**, 575–579 (1997)
36. D. Boschi, S. Branca, F. De Martini, L. Hardy, S. Popescu, Experimental realization of teleporting an unknown pure quantum state via dual classical and Einstein-Podolsky-Rosen channels. Phys. Rev. Lett. **80**(6), 1121–1125 (1998)
37. P. Walther, K.J. Resch, T. Rudolph, E. Schenck, H. Weinfurter, V. Vedral, M. Aspelmeyer, A. Zeilinger, Experimental one-way quantum computing. Nature **434**(7030), 169–176 (2005)
38. C.-Y. Lu, X.-Q. Zhou, O. Gühne, W.-B. Gao, J. Zhang, Z.-S. Yuan, A. Gobel, T. Yang, J.W. Pan, Experimental entanglement of six photons in graph states. Nat. Phys. **3**, 91–95 (2007)
39. A.S. Clark, J. Fulconis, J.G. Rarity, W.J. Wadsworth, J.L. O'Brien, All-optical-fiber polarization-based quantum logic gate. Phys. Rev. A **79**, 030303(R) (2009)
40. T. Honjo, K. Inoue, H. Takahashi, Differential-phase-shift quantum key distribution experiment with aplanar light-wave circuit Mach-Zehnder interferometer. Opt. Lett. **29**(23), 2797–2799 (2004)
41. H. Takesue, K. Inoue, Generation of $1.5 - \mu$m band time-bin entanglement using spontaneous fiber four-wave mixing and planar light-wave circuit interferometers. Phys. Rev. A **72**(4), 041804 (2005)
42. A. Politi, M.J. Cryan, J.G. Rarity, S. Yu, J.L. O'Brien, Silica-on-silicon waveguide quantum circuits. Science **320**(5876), 646–649 (2008)
43. G.D. Marshall, A. Politi, J.C.F. Matthews, P. Dekker, M. Ams, M.J. Withford, J.L. O'Brien, Laser written waveguide photonic quantum circuits. Opt. Express **17**(15), 12546–12554 (2009)
44. A. Politi, J.C.F. Matthews, J.L. O'Brien, Shor's quantum factoring algorithm on a photonic chip. Science **325**(5945), 1221–1221 (2009)
45. A. Laing, A. Peruzzo, A. Politi, M. Rodas Verde, M. Halder, T.C. Ralph, M.G. Thompson, J.L. O'Brien, High-fidelity operation of quantum photonic circuits. Appl. Phys. Lett. **97**, 211109 (2010)
46. J.C.F. Matthews, A. Politi, A. Stefanov, J.L. O'Brien, Manipulation of multiphoton entanglement in waveguide quantum circuits. Nat. Photonics **3**, 346–350 (2009)
47. B.J. Smith, D. Kundys, N. Thomas-Peter, P.G.R. Smith, I.A. Walmsley, Phase-controlled integrated photonic quantum circuits. Opt. Express **17**, 13516–13525 (2009)
48. J.C.F. Matthews, A. Politi, D. Bonneau, J.L. O'Brien, Heralding two-photon and four-photon path entanglement on a chip. Phys. Rev. Lett. **107**(163602) (2011)
49. L. Sansoni, F. Sciarrino, G. Vallone, P. Mataloni, A. Crespi, R. Ramponi, R. Osellame, Polarization entangled state measurement on a chip. Phys. Rev. Lett. **105**, 200503 (2010)
50. A. Crespi, R. Ramponi, R. Osellame, L. Sansoni, I. Bongioanni, F. Sciarrino, G. Vallone, P. Mataloni, Integrated photonic quantum gates for polarization qubits. Nat. Commun. **2**(566) (2011)
51. A. Peruzzo, A. Laing, A. Politi, R. Rudolph, J.L. O'Brien, Multimode quantum interference of photons in multiport integrated devices. Nat. Commun. **2**(224) (2011)

52. H.B. Perets, Y. Lahini, F. Pozzi, M. Sorel, R. Morandotti, Y. Silberberg, Realization of quantum walks with negligible decoherence in waveguide lattices. Phys. Rev. Lett. **100**(17), 170506 (2008)
53. E. Keil, A. Szameit, F. Dreisow, M. Heinrich, S. Nolte, A. Tünnermann, Photon correlations in two-dimensional waveguide arrays and their classical estimate. Phys. Rev. A **81**, 023834 (2010)
54. A. Peruzzo, M. Lobino, J.C.F. Matthews, N. Matsuda, A. Politi, K. Poulios, X.-Q. Zhou, Y. Lahini, N. Ismail, K. Wörhoff, Y. Bromberg, Y. Silberberg, M.G. Thompson, J.L. O'Brien, Quantum walks of correlated photons. Science **329**, 1500–1503 (2010)
55. J.O. Owens, M.A. Broome, D.N. Biggerstaff, M.E. Coggin, A. Fedrizzi, T. Linjordet, M. Ams, G.D. Marshall, J. Twamley, M.J. Withford, A.G. White, Two-photon quantum walks in an elliptical direct-write waveguide array. New J. Phys. **13**, 075003 (2011)
56. J.C.F. Matthews, K. Poulios, J.D.A. Meinecke, A. Politi, A. Peruzzo, N. Ismail, K.Wörhoff, M.G. Thompson, J.L. O'Brien, Simulating quantum statistics with entangled photons: a continuous transition from bosons to fermions. arxiv:1106.1166 (2011)

Chapter 2
Background and Methods

2.1 Introduction

The purpose of this chapter is to collect experimental details and background material that are common to multiple chapters of this thesis. Section 2.2 discusses a model of spontaneous parametric down conversion and details various experimental down conversion setups used for multi-photon state generation: each description will refer to the chapters where they are employed. Section 2.3 reviews the underlying concepts of waveguide optics required for this thesis. Finally, Sect. 2.4 describes three different waveguide architectures that are employed in the experiments reported in this thesis. Simulation and fabrication of these architectures lies outside the scope of this thesis, and were performed by colleagues as stated in each corresponding section.

2.2 Spontaneous Parametric Down Conversion

A range of single photon sources exist. Attenuated lasers can approximate a single photon source, allowing proof of principle experiments requiring only single photonic qubits—for example quantum cryptography [1]—but are not suitable for producing more than one photon for quantum interference due to cross terms in the tensor product of two coherent states [2]. An ideal source of photons for scalable quantum technology is a truly "on-demand" source that deterministically initiates emission of one photon from either an atomic or an atom-like system (such as quantum dots or diamond colour centres). Ongoing research is currently developing two or more identical single-photon emitters and has included demonstrations of quantum interference effects using quantum dots [3, 4], trapped ions [5, 6] and trapped atoms [7, 8]. However, currently single photon emitters are yet to deliver high visibility two photon quantum interference while quantum interference of more than two photons from single emitters is yet to be demonstrated.

J. C. F. Matthews, *Multi-Photon Quantum Information Science and Technology in Integrated Optics*, Springer Theses, DOI: 10.1007/978-3-642-32870-1_2,
© Springer-Verlag Berlin Heidelberg 2013

In the absence of fully developed on-demand multi-photon sources, we turn to post-selected multi-photon states arising from *spontaneous parametric down conversion* (SPDC) which is currently the favoured method used to produce degenerate photon pairs for quantum interference in linear optics and the process used to generate photons for all experiments reported in this thesis. Since the foundational quantum interference experiments using SPDC [9–11] there have been a vast number of quantum interference applications for down conversion including fundamental tests of quantum physics (e.g. non-locality [12, 13] and quantum teleportation [14]), proof of principle quantum computing (e.g. two-qubit gates [15], compiled quantum algorithms [16, 17] and quantum chemistry [18]) and quantum metrology (e.g. for observing of two-[19] three-[20] four-[21, 22] and five-photon [23] number-path entanglement dynamics).

SPDC generates pairs of low energy photons from an initial pump laser beam focused in a non-linear medium and is either a four-wave mixing process, as reported for example with photonic crystal fibre sources (e.g. [24]), or a three-wave mixing process. In the latter case, SPDC is understood to be the generation of two lower energy photons (referred to as the signal and idler photons) from a photon in the pump field, mediated by a material with a nonzero χ^2 value. Conservation of energy dictates the frequency of the signal and idler photons must sum to that of the pump photon ($\omega_p = \omega_s + \omega_i$). Furthermore, the momentum vectors of the signal and idler photons must sum to that of the pump photon in order for momentum to be conserved ($\mathbf{k}_p = \mathbf{k}_s + \mathbf{k}_i$).

The approximate quantum state resulting from type I SPDC is computed using the interaction Hamiltonian

$$H_I = i\hbar\chi a_s^\dagger a_i^\dagger - i\hbar\chi^* a_s a_i \tag{2.1}$$

for bosonic creation operators $a_s^\dagger$ and $a_i^\dagger$ acting on the signal s and idler i fields. Here we have adopted the approximation that the pump is a classical field whose amplitude is incorporated in the parameter χ [25]. At this point we note that SPDC can be modelled to excite different polarisation modes which categorises SPDC as being type I—where the generated photons have the same polarisation—or type II—where the polarisation of generated signal and idler photons are orthogonal. All experiments reported in this thesis use type I SPDC.

Unitary evolution of the Hamiltonian acting on the initial state of vacuum in the signal and idler modes $|\psi(0)\rangle = |0\rangle_s |0\rangle_i$ is given by

$$U(t)|\psi(0)\rangle = e^{-iH_I t/\hbar}|\psi(0)\rangle = \exp\left\{\chi t a_s^\dagger a_i^\dagger - \chi^* t a_s a_i\right\}|0\rangle_s |0\rangle_i \tag{2.2}$$

$$\approx \exp\{\xi a_s^\dagger a_i^\dagger\}|0\rangle_s |0\rangle_i \tag{2.3}$$

$$= \sum_{k=0}^\infty \frac{\xi^k}{k!} a_s^{\dagger k} a_i^{\dagger k}|0\rangle_s |0\rangle_i \tag{2.4}$$

$$= |0\rangle_s |0\rangle_i + \xi |1\rangle_s |1\rangle_i + \xi^2 |2\rangle_s |2\rangle_i + \xi^3 |3\rangle_s |3\rangle_i + \ldots \tag{2.5}$$

where $\xi \equiv \chi t$ is a parameter dependent upon the interaction time t, the strength of the nonlinearity in the material and the power in the pump beam. The approximation in line 2.3 uses a normal ordering step [25] and assumes the regime of $|\xi| \ll 1$.

Direct control of the laser pump allows control over the probability amplitudes of each term in the SPDC state in Eq. (2.5). Choosing small ξ such that ξ^2 and higher order terms become negligible with respect to ξ provides an excellent approximation to a two photon state in superposition with the vacuum:

$$|0\rangle + \xi |1\rangle_s |1\rangle_i \qquad (2.6)$$

Recording twofold coincidental detection events at the output from the source (or after some optical circuit), post-selects only the single pair state $|1\rangle_s |1\rangle_i$ as the input. This is the approach used for experiments reported in Chaps. 3, 4, 5, 7 and 8; experimental details of a continuous-wave pumped two photon down conversion source are given in Sect. 2.2.1.

Increasing the pump power increases the amplitude of states associated to more than one pair of photons, such as the four-photon term $\xi^2 |2\rangle_s |2\rangle_i$. However, increasing ξ also increases the rate of single pairs, which is problematic for a post selection technique that cannot resolve coherently generated multi-pair states and multiple pairs of photons that are temporally distinguishable. Let us consider the following example. Suppose we have a finite coherence of photons generated—which in our case is defined spectrally via filtering ($\lambda \approx 800\,\mathrm{nm}\ \Delta\lambda \approx 3\,\mathrm{nm}$) and is typically of the order 10^{-12} s. A photon pair generated in one window of time τ will be distinguishable from a second pair generated outside the coherence time of the first pair, in a second window τ'. This means two photon pairs in the state $|1\rangle_{s,\tau} |1\rangle_{t,\tau} |1\rangle_{s,\tau'} |1\rangle_{t,\tau'}$ will quantum-interfere in a different manner, in some optical circuit, to two photon pairs produced coherently together in the same time bin in the state $|2\rangle_{s,\tau} |2\rangle_{t,\tau}$. Successful post-selection of photons that have been generated within a time-frame equal to the coherence time of the generated photons would require a detector timing resolution of order 10^{-12}s, compared to the $O(10^{-9})$ s jitter for silicon avalanche photo-diode single photon counting modules.

For experiments requiring the state $|n\rangle_s |n\rangle_i$ for $n > 1$, an effective gating can be achieved using a pulsed laser to pump the SPDC process [25]. Making the pulse length sufficiently short ($\sim 100\,\mathrm{fs}$) ensures down converted photons can only be produced within time-frames comparable to the coherence time of the generated photons. Separating each pump pulse by more than the detector jitter and longer than the coincidence window of the counting electronics, implies coincidental detection events arise from photons generated by a single pulse from the laser pulse. Photon counting is then used in post-selection to ignore quantum interference effects of lower photon number terms; for example counting fourfold coincidence events ignores the $|0\rangle$ and $|1\rangle_s |1\rangle_i$ terms.

Of course, increasing the laser power, such that the $\xi^2 |2\rangle_s |2\rangle_i$ becomes an appreciable event, allows higher order terms to become non-negligible. This highlights one of the major limiting factors for multi-photon experiments based on single crystal down conversion. In the absence of number resolving detectors, higher order terms become a source of noise. For example, four photons from the six photon state $|3\rangle_s |3\rangle_i$ can be detected and counted as a data-point in a four-photon experiment. One approach is to choose a level of noise that is acceptable to the experiment and set ξ accordingly. For example in a simplified lossless case, setting the single pair generation probability $|\xi|^2 \approx 0.1$ implies an $n+2$ photon term has approximately ten times less probability amplitude than the n photon term. However each experiment must be considered individually with respect to the details of the detection scheme and the overall lumped detection efficiency of the experiment. Pulsed SPDC is used in the multi-photon experiments reported in Chaps. 3, 5 and 6. Details of the photon sources used for those experiments are given in Sect. 2.2.2.

2.2.1 Degenerate Photon Pair Sources

The two-photon experiments reported in this thesis are based on the degenerate photon pair source presented in Fig. 2.1. A vertically polarised 404 nm continuous wave (CW), diode laser (Toptica iBeam) is focused (minimum beam waist $\sim$40 μm) with 60 mW of power on a 2 mm thick Type I phase matched Bismuth Borate BiB_3O_6 (BiBO). The phase matching is such that photon pairs are generated non-collinear, with signal and idler photons emitted in separate spatial modes for collection into optical fibres. Conservation of momentum dictates photon pairs of equal frequency $\omega_s = \omega_i = \omega_p/2$ are emitted in a cone structure that is symmetric around the initial pump beam, as indicated in Fig. 2.1a: this can be used to collect a single pair of photons in two spatial modes (Fig. 2.1), or to collect single pair of photons in two of four spatial modes (Fig. 2.3). The desired degenerate photons are filtered using high transmission ($>$95 %) laser line interference filters (Semrock) with centre wavelength ($\lambda_0 = 808$ nm) matched to the degenerate photon pairs $\lambda_s = \lambda_i$—fine tuning of wavelength below λ_0 can be achieved by tilting the filter with respect to the incoming photons.

The photons are then collected using mirrors and aspheric lenses as shown in Fig. 2.1b into polarisation maintaining fibre—the axis of each fibre is aligned to the horizontal plane. Alignment of the photon source begins with a coarse alignment (using a visible alignment laser fed back through the fibres onto the centre of the BiBO crystal) and a fine alignment based on first maximising single photon count rates detected in each mode and then on coincidental events across the two output modes. The degeneracy of the photons is then tested via a Hong–Ou–Mandel experiment (see Chap. 3). Typical two photon count rates straight from the source as described— detected with two $\sim$70 % efficient single photon counting modules (PerkinElmer AQR family)—is of the order 50 kHz. Typical collection efficiency of each mode is $\sim$20 %.[1]

---

[1] Where we define collection efficiency of one mode A of two modes A and B, as the total number coincidences detected from modes A and B, divided by the total number of single photon events detected from mode A; note that this definition involves the detector efficiency.

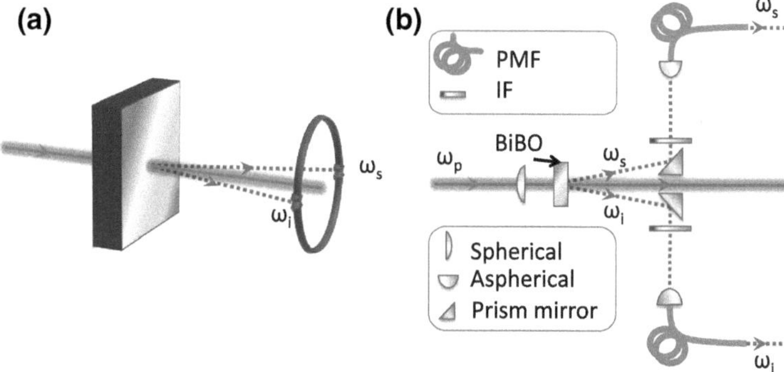

Fig. 2.1 Non-colinear SPDC generates degenerate photon pairs symmetrically in a cone structure around the initial pump. This can be used to collect a single pair of photons in two spatial modes (**a**) using prism mirrors and polarisation maintaining fibre (PMF) (**b**). Interference filters (IF) are used to ensure the desired degenerate photon pairs are collected

We also note that the polarisation state of the degenerate photon pairs emitted in the SPDC process can be manipulated with bulk optical wave plates. This is reported in more detail in Chap. 8 for post selection of polarisation entanglement in a waveguide circuit.

2.2.2 Multi-Photon Sources

The four-photon experiments reported in Chaps. 3 and 5 and the four- and six-photon experiments reported in Chap. 6 were conducted using degenerate single photon pairs produced via SPDC pumped by a pulsed laser source. Photons used in the experiment reported in Chap. 4 were generated with the same multi-photon source setup, operated with low laser pump power to approximate a two-photon source. A schematic of the typical setup is given in Fig. 2.2. The four photon states for experiments in Chaps. 3 and 5 were generated at the wavelength $\lambda_s = \lambda_i = 780$ nm and filtered with high transmission interference filters ($>95\,\%$, Semrock, $\lambda_0 = 780$ nm, $\Delta\lambda = 3$ nm). The four and six photon states for experiments reported in Chap. 6 were generated with at the wavelength $\lambda_s = \lambda_i = 785$ nm and filtered with high transmission interference filters (Semrock, $>95\,\%$, $\lambda_0 = 785$ nm, $\Delta\lambda = 3$ nm).

The nonlinear crystal used for SPDC was a 2 mm thick, Type-I phase matched Bismuth Borate BiB_3O_6 (BiBO) pumped by a pulsed (blue) $\frac{1}{2}\lambda_s$ nm beam, focused to a waist of $\omega_0 \approx 40\,\mu$m. The blue pump was prepared using a further 2 mm thick BiBO crystal, phase matched for second harmonic generation (SHG) to double the frequency of a mode-locked 80 MHz repetition rate, 150 fs pulse length Ti:Sapphire laser (Coherent Chameleon Ultra II) focused to a waist of $\omega_0 \approx 40\,\mu$m; four successive dichroic mirrors (DM) are used to purify the pump beam spectrally. Degenerate

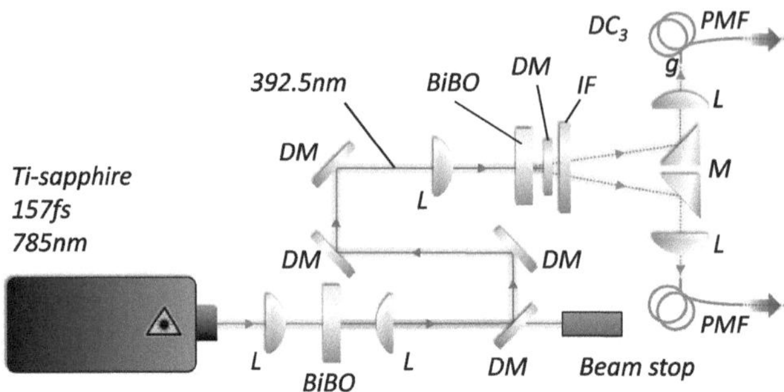

Fig. 2.2 Schematic of the two-mode pulsed SPDC multi-photon source used to generate the states $|n\rangle_s\,|n\rangle_i$ via post selection, for $n = 1, 2, 3$. Figure reproduced from [26]

photon pairs are created by the SPDC crystal and pass through $\Delta\lambda = 3$ nm interference filters (IF) which filter each photon to a coherence length of $l_c = \lambda^2/\Delta\lambda \approx 200$ μm. The photons are collected into two single mode polarization maintaining fibers (PMFs) coupled to two diametrically opposite points on the SPDC cone, as described in the previous sub-section. In the case of low average pump power, the state $|1\rangle_s\,|1\rangle_i$ is produced with a rate of 100 s^{-1}. On increasing the average pump power, the multi-photon production rate from the down-conversion process is no longer negligible such that two and three degenerate pairs of photons can be observed via the down conversion terms $|2\rangle_s\,|2\rangle_i$ and $|3\rangle_s\,|3\rangle_i$ respectively.

The cone-structure for emitting non-colinear photon pairs in type I down conversion allows for multi-mode collection as given in Fig. 2.3 for the case of collecting four spatial modes. This source was used for the Shor's algorithm experiment [27] reported in Chap. 4, which required two pairs of photonic qubits launched into a circuit that only required quantum interference between the signal and idler photons of each pair. Due to the requirements of the circuit, photons were generated at $\lambda_s = \lambda_i = 790$ nm using an appropriate pump ($\lambda_p = 395$ nm) and interference filters ($\lambda_0 = 790$ nm, $\Delta\lambda = 3$ nm).

Detection of multiple photon states in the same optical mode is accomplished non-deterministically using cascaded non-number resolving, optical fibre-coupled single photon counting modules (see Chaps. 3, 5 and 6 for relevant details).

2.3 Single Waveguide Propagation

Optical waveguides are the main structure employed throughout this thesis for linear optical circuits, and are a thoroughly understood technology, widely used for optical telecommunications in the form of optical fibres and monolithic integrated waveguide chips. Here we present some fundamental descriptions of optical waveguides relevant to the experiments presented in this thesis.

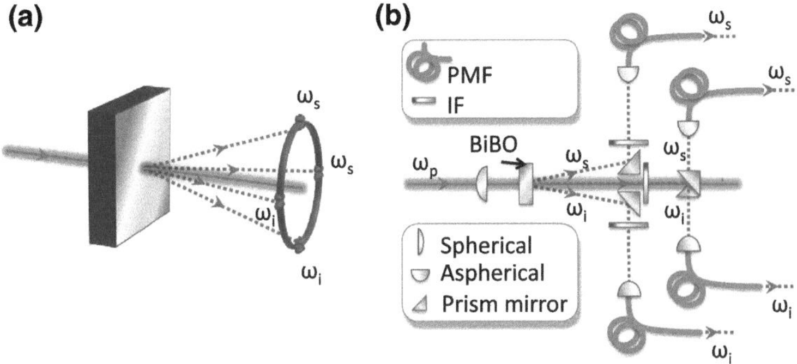

Fig. 2.3 The spontaneous parametric down conversion setup for collecting photon pairs in four spatial modes for the Shor's algorithm experiment [27] reported in Chap. 4

Light confining structures can be classified according to the number of spatial dimensions of confinement: planar waveguides or slab waveguides confine light in only one dimension; channel waveguides and optical fibres confine light in two orthogonal directions; photonic crystal structures, for example, can confine light in one-, two- or three-dimensions, depending on the structure. Channel waveguides are the structures on which our quantum optical circuits are based and the modelling of which requires numerical solution of the Maxwell's equations using commercially available beam propagation software. However, to understand some of the physics of integrated optics, in the next few sections we consider the more simple model of one-dimensional confinement in a step index waveguide, which will provide an insight into the underlying behaviour of waveguide propagation.

2.3.1 Ray Optics Description: Total Internal Reflection

There are several requirements for waveguided optics. First of all, the guiding structure must consist of a core with refractive index n_1 and cladding with lower refractive index n_0 (such that $n_1 > n_0$), as indicated in the refractive index profile in Fig. 2.4b. In a ray model of light propagation total internal reflection (TIR), illustrated in Fig. 2.4a, ensures light coupled to the end facet of a waveguide structure is confined to the core, provided the condition for TIR is satisfied:

$$n_1 \sin (\pi/2 - \phi) \geq n_0 \tag{2.7}$$

The value of ϕ for which Eq. (2.7) is an equality is known as the critical angle. Snell's law tells us that $\sin \theta = n_1 \sin \phi$, therefore we have an upper bound $\theta_{\max}$ (known as the numerical aperture) on the incident angle θ that will satisfy the TIR condition:

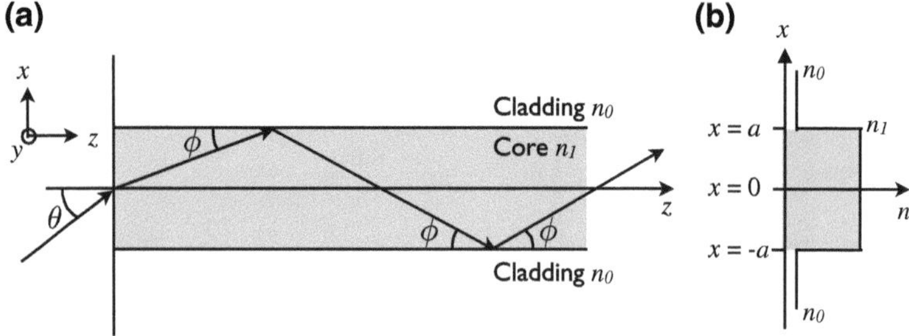

Fig. 2.4 **a** In the ray picture, light is confined in a planar waveguide by total internal reflection. **b** Illustrates the refractive index (n) profile. Figure inspired by [28]

$$\theta \leq \sin^{-1} \sqrt{n_1^2 - n_0^2} \equiv \theta_{\max} \tag{2.8}$$

TIR is a necessary, but not sufficient for guiding light in a waveguide structure. Even if ϕ is smaller than the critical angle for TIR, arbitrary values of ϕ are not permitted for light guidance. By considering phase fronts of the rays and phase changes on reflection (of the core-cladding interfaces) leads to a description of permitted solutions of ϕ_m, associated to integer values $m = 0, 1, 2, \ldots$ (see [28] for example). These solutions are equivalent to the permitted guided modes that are derived from an electro-magnetic description (described in the following section). Indeed, there is good agreement between describing waveguides with ray optics and with electromagnetism. However, the simple ray optics description considered here fails in describing evanescent fields of light that lay outside of the core—a key mechanism for coupling light from one waveguide to another.

2.3.2 An Electromagnetic Description: Guided Modes and Evanescent Fields

The purpose of this section is to highlight two key features of the electromagnetic description of waveguides. The first will be that guided electro-magnetic fields have an evanescent field that lies outside of the core of a waveguide, which allows coupling between multiple waveguides and other structures. The second point is that there are discrete solutions to how light can propagate in a wave guiding structure, known as modes. As with the previous section, we shall make these two points by studying the simple model of one-dimensional confinement; the more complex situation of two-dimensional confinement of channel waveguides is solved numerically using commercial software.

With the condition that light is propagating through a dielectric, non-magnetic, isotropic and linear medium, Maxwell's equations simplify to

$$\nabla \cdot \mathcal{E} = 0; \ \nabla \cdot \mathcal{H} = 0; \ \nabla \times \mathcal{E} = -\mu_0 \frac{\partial \mathcal{H}}{\partial t}; \ \nabla \times \mathcal{H} = \varepsilon_0 n^2 \mu_0 \frac{\partial \mathcal{E}}{\partial t} \qquad (2.9)$$

where μ_0 is the free space permeability, ε_0 is the free space permittivity $\mathcal{E}$ and $\mathcal{H}$ are the electric and magnetic fields respectively. By definition waveguides are inhomogeneous, meaning the refractive index is position dependent: $n = n(\mathbf{r})$. The inhomogeneous wave equations [29] derived from Maxwell's equations—assuming the structure is isotropic, linear, dielectric and non-magnetic—take the form

$$\nabla^2 \mathcal{E} + \nabla \left(\frac{1}{n(\mathbf{r})^2} \nabla n(\mathbf{r})^2 \mathcal{E} \right) - \varepsilon_0 \mu_0 n(\mathbf{r})^2 \frac{\partial^2 \mathcal{E}}{\partial t^2} = 0 \qquad (2.10)$$

$$\nabla^2 \mathcal{H} + \frac{1}{n(\mathbf{r})^2} \nabla n(\mathbf{r})^2 \times (\nabla \times \mathcal{H}) - \varepsilon_0 \mu_0 n(\mathbf{r})^2 \frac{\partial^2 \mathcal{H}}{\partial t^2} = 0 \qquad (2.11)$$

For the step index situation described in Fig. 2.4b, the refractive index in the core, and the cladding is assumed to be respectively constant, therefore the second term for each wave equation are zero.

For plane wave propagation in two-dimensional confinement in the z direction, the solutions to the wave equations are

$$\mathcal{E}(\mathbf{r}, t) = \mathbf{E}(x, y) e^{i(\omega t - \beta z)} \qquad (2.12)$$

$$\mathcal{H}(\mathbf{r}, t) = \mathbf{H}(x, y) e^{i(\omega t - \beta z)} \qquad (2.13)$$

where β is the propagation constant of the wave in the structure and ω is the angular frequency of the wave. This is simplified further for one-dimensional confinement in Fig. 2.4a by realising the wave has no dependence on the y direction: $\partial E_y / \partial y = 0 = \partial H / \partial y$.

To compute the form of permitted propagation, the two separate situations of the transverse electric (TE) and the transverse magnetic (TM) waves are treated. TE propagation are defined from electric field components that are perpendicular (transverse) to the plane of incidence (Fig. 2.4 sketches the incident plane as the $x - z$). The TM propagation is defined to be comprised of electric field components that are parallel to the plane of incidence and therefore magnetic field components that are transverse to the plane of incidence. Since the derivation of the TE and TM propagation follows a similar treatment, we shall only be concerned with TE fields here.

For TE propagation only E_y, H_x and H_z are non-zero. Subject to these conditions, substituting the planar wave equation solutions Eqs. (2.12), (2.13) into the wave equations Eqs. (2.10), (2.11) yield a second order differential equation involving E_y

$$\frac{d^2 E_y}{dx^2} + \left[k_0 n^2(x) - \beta^2 \right] E_y = 0 \qquad (2.14)$$

where $n(x)$ is defined by the refractive index profile in Fig. 2.4b, k_0 is related to the wavelength λ_0 of the light in free space and the angular frequency ω by $k_0 = 2\pi/\lambda_0$, $\omega = 2\pi c/\lambda_0$ and $c = 1/\sqrt{\varepsilon_0\mu_0}$.

For guided mode, the propagation constant β must satisfy [29]

$$k_0 n_0 < \beta < k_0 n_1 \tag{2.15}$$

So for the core and cladding regions, we re-write Eq. (2.14) as

$$\frac{\partial^2 E_y}{\partial x^2} - \gamma^2 E_y = 0 \quad x \leq -a, \text{ or } x \geq a \tag{2.16}$$

$$\frac{\partial^2 E_y}{\partial x^2} + \kappa^2 E_y = 0 \quad -a < x < a \tag{2.17}$$

where we have defined the real parameters

$$\gamma^2 = \beta^2 - k_0^2 n_0^2; \quad \kappa^2 = k_0^2 n_1^2 - \beta^2 \tag{2.18}$$

The solution of Eq. (2.14) therefore gives the electric field of guided TE propagation:

$$E_y = \begin{cases} A e^{\gamma x} & x \leq -a \\ B e^{i\kappa x} + C e^{-i\kappa x} & -a < x < a \\ D e^{-\gamma x} & x \geq a \end{cases} \tag{2.19}$$

For some constants A, B, C, D, γ and κ that are dependent upon the waveguide parameters (a, n_0 and n_1), the propagation constant β and the free-space wavelength λ_0. Following a similar treatment, the TM field has a similar solution.

Examining Eq. (2.19), we arrive at our first point: The guided TE (and similarly the guided TM) field has part of its solution outside of the waveguide core in the form of an exponentially decreasing profile for $x \leq -a$ and $x \geq a$. This corresponds to the *evanescent field*, which is the mechanism that allows light to couple from one waveguide into either neighbouring waveguides (or other structures), when the core of the neighbouring waveguide intersects the evanescent field of the original. We shall look at the behaviour of two coupled waveguides in the next section.

Returning to the solution for the mode profile in Eq. (2.19), boundary conditions dictate continuity of the quantities E_y and $\partial E_y/\partial x$ at the interfaces between core and cladding ($x = -a$ and $x = a$), leading to four equations of the five unknown parameters β, A, B, C and D. Solving this system [29] leads to the so called dispersion relation

$$\tan 2a\kappa = \frac{2\gamma/\kappa}{1 - \gamma^2/\kappa^2} \tag{2.20}$$

Tangent is a periodic modulo π, therefore we have a potentially infinite set of solutions governed by

$$\tan 2\kappa a = \tan(2\kappa a + m\pi), \quad m = 0, 1, 2, \ldots \tag{2.21}$$

Here we arrive at a second point of interest: The TE (and TM) field as a discrete set of permitted solutions, characterised by m, known as *modes*. Since the dispersion relation 2.20 has a dependency on the properties of the waveguide and of the light, it follows that the number of permitted modes shares the same dependency: for the case that there is only one solution ($m = 0$) the waveguide is defined to be *single mode*, for $m > 0$, the waveguide is referred to as *multi-mode*. For the purposes of all experiments reported in this thesis, we required single mode operation and the waveguides used were designed accordingly.

2.3.3 Coupled Waveguides: A Quantum Mechanical Description

We have seen in the previous section that waveguides have an evanescent field that allow coupling between neighbouring waveguides. This forms the basis for directional couplers, where two waveguides are close enough to interact with each others evanescent field, and is one of the simplest methods to create beamspiltter devices in an optical waveguide. In classical optics, this is often modelled with an electromagnetic description which can be found in a multitude of classical integrated optics textbooks.[2] Instead, here we take the more direct approach of modelling the system with a Hamiltonian, acting on single photon Fock states.

We saw that the evanescent field is exponentially decreasing away from the core-cladding interface, therefore we can model two identical, uniform evanescently coupled waveguides (labeled 1 and 2) as two nearest neighbour coupled oscillators [30], with the Hamiltonian[3]

$$H = \beta a_1^\dagger a_1 + \beta a_2^\dagger a_2 + C a_2^\dagger a_1 + C a_1^\dagger a_2 \tag{2.22}$$

for propagation constants β and coupling strengths proportional to constants C dependent upon the waveguide structure and the wavelength. Here we are assuming each waveguide is single mode, and we consider H as a transformation acting on the two waveguides with the two dimensional vectors $(1, 0)^T$ and $(0, 1)^T$ respectively corresponding to the single modes of waveguide 1 and waveguide 2. Coupling is therefore represented in this basis by the matrix

$$H \doteq \begin{pmatrix} \beta & C \\ C & \beta \end{pmatrix} \tag{2.23}$$

Unitary evolution along the propagation direction z is therefore described by computing the 2×2 matrix

---

[2] For example Refs. [28, 29].

[3] This approach is also used for theoretical treatment of larger systems of coupled waveguides (for example [31, 32]), which we shall return to in Chap. 7.

$$e^{iHz} \doteq \frac{1}{\sqrt{2}} \begin{pmatrix} 1 & 1 \\ 1 & -1 \end{pmatrix} \begin{pmatrix} \beta & C \\ C & \beta \end{pmatrix} \frac{1}{\sqrt{2}} \begin{pmatrix} 1 & 1 \\ 1 & -1 \end{pmatrix} \qquad (2.24)$$

$$= e^{iz\beta} \begin{pmatrix} \cos(zC) & i\sin(zC) \\ i\sin(zC) & \cos(zC) \end{pmatrix} \qquad (2.25)$$

Interaction between the two waveguides is controlled by bending the waveguides into the evanescent coupling regime and then subsequently separating them after some interaction region z. In this manner, an appropriate reflectivity beamsplitter can be created with two waveguides by controlling z and C—dependent upon properties of each waveguide, the separation of the two waveguides and of the guided light. To see this, e^{iHz} acting on $(1, 0)^T$ for example, we see that light initially present in waveguide 1 couples sinusoidally with z to waveguide 2 with intensity $\sin^2 zC$. This is the same conclusion for modelling two coupled waveguides with an electromagnetic description.

2.4 Waveguide Architectures

Here, we list the key features of the architectures used for the quantum photonic experiments reported in this thesis. Three architectures were used: (i) directly laser written waveguides in silica, (ii) lithographically fabricated silica-on-silicon and (iii) lithographically fabricated silicon-oxynitride, all of which were designed for single mode operation for near infared photons for which our down conversion sources emit photons (780–810 nm) and for which commercially available silicon based avalanche photo-diodes can detect photons. Furthermore, all waveguides used were designed to have a small amount of birefringence (for example, the silicon oxynitride waveguides have a total birefringence is of the order $\Delta n = n_{\mathrm{TM}} - n_{\mathrm{TE}} = 2 * 10^{-4}$). Small values of birefringence provides polarization preserving propagation for the TE and TM photons throughout the device, while ensuring a nearly identical mode overlap, and coupling ratio, between different waveguides in the coupled array for TM and TE modes: this is of paramount importance for the experiment reported in Chap. 8. In general, preserving polarisation states of degenerate photons throughout optical circuitry is important for high quality quantum interference and for encoding quantum information.

2.4.1 Directly Written Waveguides in Silica

Directly written waveguides are fabricated with a tightly focused femtosecond laser machining process to locally alter the refractive index inside glass to inscribe waveguide circuits (Fig. 2.5a). Using nano-positioning three-axis stages, waveguides can be written in three-dimensional architectures, without the need for lithography, yielding a technique that can offer rapid prototyping of "one-off" circuits and architectures that cannot be realised in planar architectures allowing, for example,

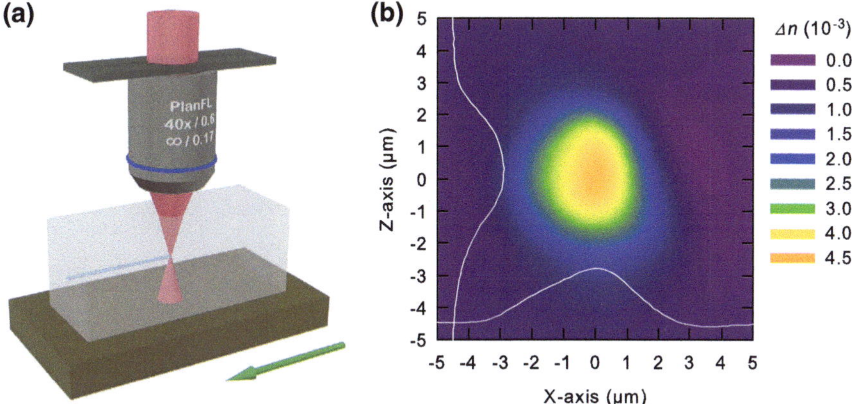

Fig. 2.5 Directly laser written waveguides. **a** Cartoon of a focused femotosecond laser writing a waveguide in pure silica. **b** Refractive index profile (Δn) of a directly written waveguide, measured by colleagues from Macquarie University. Figure reproduced from [33]

integration with micro-fluidics. Furthermore, the direct-write technique can fabricate devices with circular mode profiles, the size and ellipticity of which can be altered by controlling the laser focusing conditions. This is of particular use for producing waveguides that better match fibre modes for example, reducing optical loss.

Our collaborators at Macquarie University used this technique [34, 35] to fabricate two chips of high purity silica with a number of direct-write quantum circuits (DWQCs) composed of 2×2 directional couplers (Fig. 2.6) and Mach Zehnder interferometers for the experiments reported in Chap. 3 and Ref. [33]. The writing process created circular waveguides with an approximately Gaussian shaped refractive index profile that supported a single transverse mode with orthogonal $1/e^2$ widths of $6\,\mu$m $\times$ $6\,\mu$m measured at 806 nm. The design of directional couplers was functionally identical except for the length of the central evanescently coupled region that was varied to achieve different coupling ratios. The curved regions of the waveguides (maximum bend radius $\sim$15 mm) were of raised-sine form and connected the input and output waveguide pitch of $250\,\mu$m down to the closely spaced $10\,\mu$m evanescent coupling region of the waveguides. When butt-coupled to arrays of optical fibre with refractive index matching liquid, a typically 70 % transmission from facet to facet was routinely achieved.

2.4.2 Silica-on-Silicon

The second architecture used was lithographically fabricated doped silica waveguides on a silicon substrate. These waveguides were designed by my colleague Alberto Politi using the numerical simulation package Beam Prop, while fabrication was outsourced to a commercial integrated photonics fabrication company (The Centre

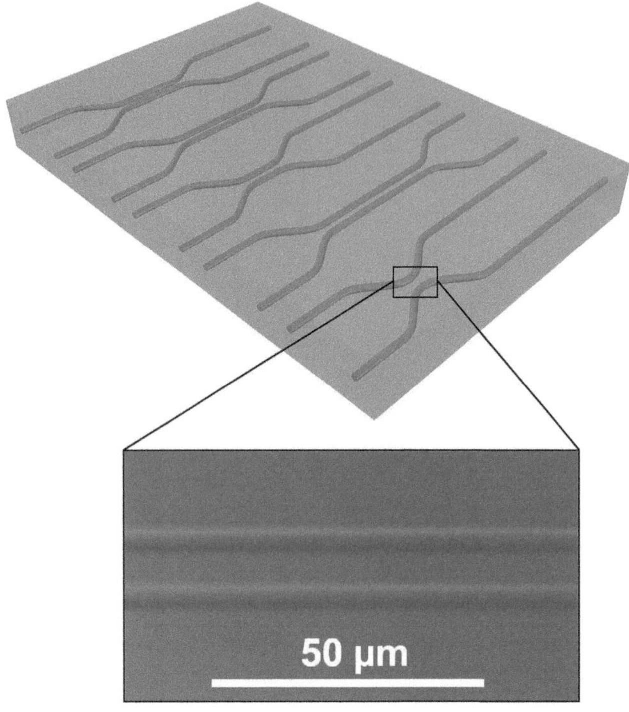

Fig. 2.6 Schematic of directional couplers fabricated by the direct-write process. The enlarged inset is an optical micrograph of the waveguides in the coupling region of one of the directional couplers, illustrating a designed 10 μm separation. Figure reproduced from [33]

Fig. 2.7 Cross-section of silica-on-silicon waveguides. Figure reproduced from [37]

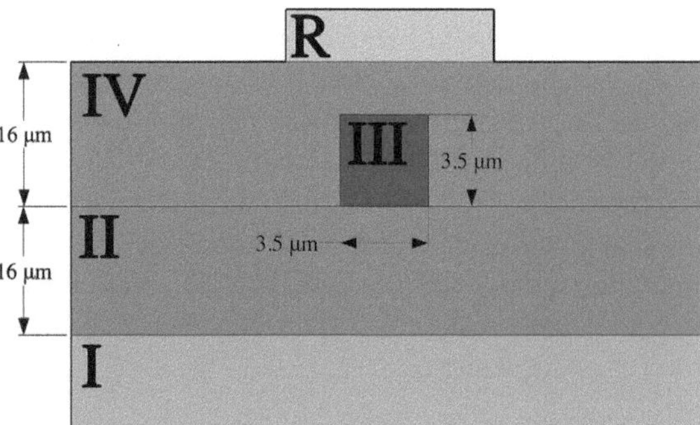

for Integrated Photonics, CIP). This architecture was used for the first quantum interference experiments in monolithic waveguide circuits [36].

The optical circuits used for experiments reported in Chaps. 4, 5 and 6 all use silica-on-silicon waveguides, illustrated in cross-section in Fig. 2.7. The waveguides were fabricated on a 4" silicon wafer (material I), onto which a 16 μm layer of thermally grown undoped silica was deposited as a buffer to form the lower cladding of the waveguides (*II*). A 3.5 μm layer of silica doped with germanium and boron oxides was then deposited by flame hydrolysis; the material of this layer constitutes the core of the structure and was patterned into 3.5 μm wide waveguides via standard optical lithographic techniques (*III*). The 16 μm upper cladding (*IV*) is phosphorus and boron doped silica with a refractive index matched to that of the buffer. Simulations indicated single mode operation at 780 nm and 800 nm. A refractive index contrast of $\Delta = (n_{core}^2 - n_{cladding}^2)/2n_{core}^2 = 0.5\,\%$ yields a minimum bend radius of $\sim$15 mm. A final metal layer was lithographically patterned on the top of the devices to form resistive elements (*R*) used for controlling optical phase shifts (reported in Chaps. 5 and 6) and metal contact pads. When butt-coupled to arrays of optical fibre with refractive index matching liquid, $\sim$70 % transmission from facet to facet was routinely achieved.

2.4.3 Silicon Oxynitride Waveguide Arrays

The third architecture used in this thesis is silicon oxynitride (SiON) waveguides lithographically fabricated on a silicon substrate, which were designed with numerical beam propagation software (Beam Prop) by colleagues in the University of Bristol, and fabricated by colleagues at the University of Twente.

The role of this material was to realise arrays of O(10) evanescently coupled waveguide, single mode for $\sim$800 nm, that could quickly bend the waveguides from the coupling region to a separation suitable for butt-coupling to arrays of optical fibre while realising a compact optical circuit. Silica-on-silicon is not a suitable architecture for this purpose. However, the high refractive index contrast $\Delta = (n_{core}^2 - n_{cladding}^2)/2n_{core}^2 = 4.4\,\%$ of silicon oxynitride does provide a suitable platform for doing quantum experiments using waveguide arrays, as reported in Chaps. 7 and 8.

Thin 600 nm films of SiON were deposited on a silicon substrate to create the core of the waveguide by standard lithography and reactive ion etching; a 5 μm thick layer of SiO_2 was then deposited to complete the cladding of the structure. The waveguides were 1.8 μm wide and were separated by a distance of 2.8 μm between waveguides in the coupled array to define the tunnelling rate of light between adjacent waveguides (Fig. 2.8). The waveguides at the end of the array spread out, at an equal rate of relative separation between adjacent waveguides, to a pitch 125 μm for coupling to optical fibre external to the chip. The overall coupling efficiency through the chip (from input fibre to output fibre) is $\sim$10 %, which is attributed to the mode-mismatch between circular input/output optical fibres and the rectangular waveguides.

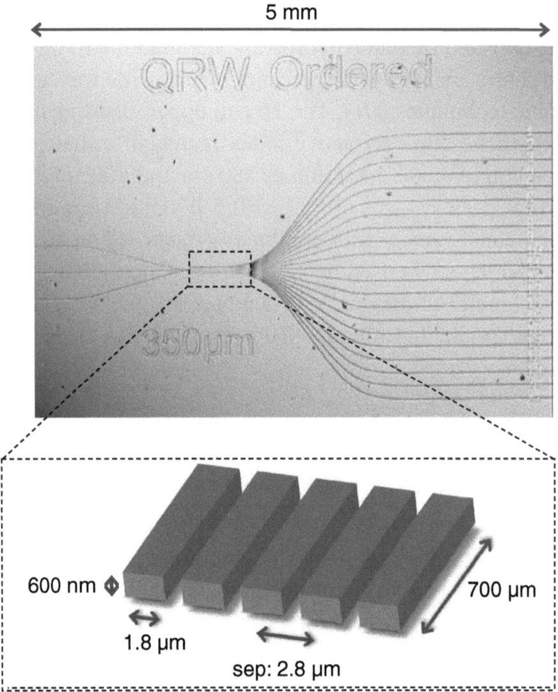

Fig. 2.8 Optical migrograph of a silicon oxynitride waveguide array chip, illustrating the coupling region (*highlighted*) and input and output bends. The enlarged section illustrates the dimensions of the waveguides and the evanescently coupled region

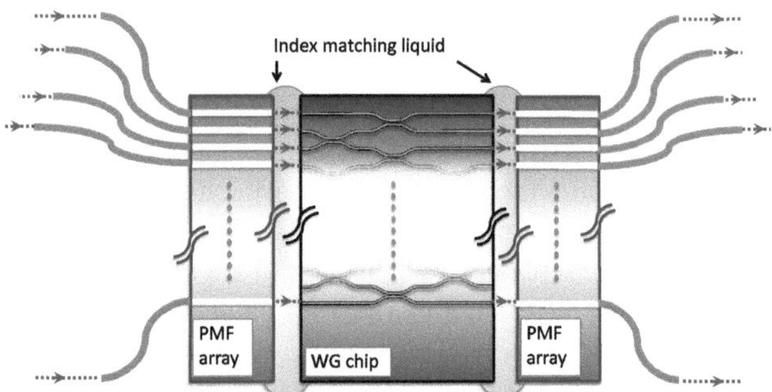

Fig. 2.9 Illustration butt-coupling arrays of optical fibre to waveguide chips. Figure reproduced from [37]

2.4.4 Chip-Fibre Coupling

For all experiments reported in this thesis, single photons were launched into, and collected from, waveguide chips using v-groove arrays of optical fibre. Each waveguide chip has a fixed spacing of input and output waveguides. For the direct-write circuits and for the silica-on-silicon circuits, this separation was $250\,\mu$m to match the standard $250\,\mu$m spacing of commercially available arrays of optical fibre (OZ optics). The input for the silicon oxynitride circuits also had a $250\,\mu$m spacing, while the output had a $125\,\mu$m spacing. Since this spacing is non standard,[4] a $250\,\mu$m array was used to access respectively all even and then all odd outputs of the array. Input fibre arrays were of polarisation maintaining fibre in all cases. The output fibre arrays used include polarisation maintaining fibre, non-polarisation maintaining single mode fibre and arrays of multimode fibre. Nano-positioning six-axis stages ($X-Y-Z$, pitch, role and yaw) were used to align the arrays to the waveguide circuits. Refractive index matching liquid was then used to reduce loss due to reflection at the facet. Fig. 2.9 illustrates a typical setup.

References

1. R. Ursin, F. Tiefenbacher, T. Schmitt-Manderbach, H. Weier, T. Scheidl, M. Lindenthal, B. Blauensteiner, T. Jennewein, J. Perdigues, P. Trojek, B. Oemer, M. Fuerst, M. Meyenburg, J. Rarity, Z. Sodnik, C. Barbieri, H. Weinfurter, A. Zeilinger, Entanglement-based quantum communication over 144 km. Nat. Phys. **3**, 481–486 (2007)
2. H.A. Bachor, T.C. Ralph, *A Guide to Experiments in Quantum Optics*, 2nd edn. (Wiley, New York, 2004)
3. C. Santori, D. Fattal, J. Vuckovic, G.S. Solomon, Y. Yamamoto, Indistinguishable photons from a single-photon device. Nature **419**(6907), 594–597 (2002)
4. R.B. Patel, A.J. Bennett, I. Farrer, C.A. Nicoll, D.A. Ritchie, A. Shields, Two-photon interference of the emission from electrically tubnable remote quantum dots. Nat. Photonics **4**, 632–635 (2010)
5. S. Olmschenk, D.N. Matsukevich, P. Maunz, D. Hayes, L.-M. Duan, C. Monroe, Quanutm teleportation between distant matter qubits. Science **323**, 486–489 (2009)
6. P. Maunz, D.L. Moerhring, S. Olmschenk, K.C. Younge, D.N. Matsukevich, C. Monroe, Quantum interference of photon pairs from two remote atomic ions. Nat. Phys. **3**, 538–541 (2007)
7. T. Chanelière, D.N. Matukevich, S.D. Jenkins, S.-Y. Lan, R. Zhao, T.A.B. Kennedy, A. Kuzmich, Quantum interference of electromagnetic fields from remote quantum memories. Phys. Rev. Lett. **98**, 113602 (2007)
8. J. Beugnon, M.P.A. Jones, J. Dingjan, B. Darqui, G. Messin, A. Browaeys, P. Grangier, Quanutm interference between two single photons emitted by independently trapped atoms. Nature **440**, 779–782 (2007)
9. C.K. Hong, Z.Y. Ou, L. Mandel, Measurement of subpicosecond time intervals between two photons by interference. Phys. Rev. Lett. **59**(18), 2044–2046 (1987)
10. J.G. Rarity, P.R. Tapster, Fourth-order interference in parametric downconversion. J. Opt. Soc. Am. B **6**, 1221–1226 (1989)
11. J.G. Rarity, P.R. Tapster, E. Jakeman, T. Larchuk, R.A. Campos, M.C. Teich, B.E.A. Saleh, Two-photon interference in a mach-zehnder interferometer. Phys. Rev. Lett. **65**(11), 1348–1351 (Sep 1990)
12. J.G. Rarity, P.R. Tapster, Two-color photons and nonlocality in fourth-order interference. Phys. Rev. A **41**, 5139–5146 (1990)
13. J.G. Rarity, P.R. Tapster, Experimental violation of bell's inequality based on phase and momentum. Phys. Rev. Lett. **64**, 2495–2498 (1990)

---

[4] The closest standard fibre-array seperation is $127\,\mu$m.

14. D. Bouwmeester, J.-W. Pan, K. Mattle, M. Eibl, H. Weinfurter, A. Zeilinger, Experimental quantum teleportation. Nature **390**, 575–579 (1997)
15. J.L. O'Brien, G.J. Pryde, A.G. White, T.C. Ralph, D. Branning, Demonstration of an all-optical quantum controlled-not gate. Nature **426**(6964), 264–267 (2003)
16. C.-Y. Lu, D.E. Browne, T. Yang, J.W. Pan, Demonstration of a compiled version of Shor's quantum factoring algorithm using photonic qubits. Phys. Rev. Lett **99**, 250504 (2007)
17. B.P. Lanyon, T.J. Weinhold, N.K. Langford, M. Barbieri, D.F.V. James, A. Gilchrist, A.G. White, Experimental demonstration of a compiled version of Shor's algorithm with quantum entanglement. Phys. Rev. Lett. **99**(25), 250505 (2007)
18. B.P. Lanyon, J.D. Whitfield, G.G. Gillett, M.E. Goggin, M.P. Almeida, I. Kassal, J.D. Biamonte, M. Mohseni, B.J. Powell, M. Barbieri, A. Aspuru-Guzik, A.G. White, Towards quantum chemistry on a quantum computer. Nat. Chem. **2**, 106–111 (2010)
19. Z.Y. Ou, X.Y. Zou, L.J. Wang, L. Mandel, Experiment on nonclassical 4th-order interference. Phys. Rev. A **42**, 2957–2965 (1990)
20. M.W. Mitchell, J.S. Lundeen, A.M. Steinberg, Super-resolving phase measurements with a multiphoton entangled state. Nature **429**(6988), 161–164 (2004)
21. P. Walther, J.W. Pan, M. Aspelmeyer, R. Ursin, S. Gasparoni, A. Zeilinger, De broglie wavelength of a non-local four-photon state. Nature **429**(6988), 158–161 (2004)
22. T. Nagata, R. Okamoto, J.L. O'brien, K. Sasaki, S. Takeuchi, Beating the standard quantum limit with four-entangled photons. Science **316**(5825), 726–729 (2007)
23. I. Afek, O. Ambar, Y. Silberberg, High-noon states by mixing quantum and classical light. Science **328**, 879–881 (2010)
24. J. Fulconis, O. Alibart, J.L. O'Brien, W.J. Wadsworth, J.G. Rarity, Nonclassical interference and entanglement generation using a photonic crystal fiber pair photon source. Phys. Rev. Lett. **99**(12), 120501 (2007)
25. Z.Y.J. Ou, *Multi-Photon Quantum Interference* (Springer, New York, 2007)
26. J.C.F. Matthews, A. Politi, A. Stefanov, J.L. O'Brien, Manipulation of multiphoton entanglement in waveguide quantum circuits. Nat. Photonics **3**, 346–350 (2009)
27. A. Politi, J.C.F. Matthews, J.L. O'Brien, Shor's quantum factoring algorithm on a photonic chip. Science **325**(5945), 1221–1221 (2009)
28. K. Okamoto, *Fundamentals of Optical Waveguides*, 2nd edn. (Elsevier Academic Press, New York, 2006)
29. G. Lifante, *Integrated Photonics: Fundamentals* (Wiley, New York, 2003)
30. L.E. Estes, T.H. Keil, L.M. Narducci, Quantum-mechanical description of two coupled harmonic oscillators. Phys. Rev. **175**(1), 286 (1968)
31. H.B. Perets, Y. Lahini, F. Pozzi, M. Sorel, R. Morandotti, Y. Silberberg, Realization of quantum walks with negligible decoherence in waveguide lattices. Phys. Rev. Lett. **100**(17), 170506 (2008)
32. Y. Bromberg, Y. Lahini, R. Morandotti, Y. Silberberg, Quantum and classical correlations in waveguide lattices. Phys. Rev. Lett. **102**(25), 253904 (2009)
33. G.D. Marshall, A. Politi, J.C.F. Matthews, P. Dekker, M. Ams, M.J. Withford, J.L. O'Brien, Laser written waveguide photonic quantum circuits. Opt. Express **17**(15), 12 546–12 554 (2009)
34. M. Ams, G.D. Marshall, D.J. Spence, M.J. Withford, Slit beam shaping method for femtosecond laser direct-write fabrication of symmetric waveguides in bulk glasses. Opt. Express **13**(15), 5676–5681 (2005)
35. M. Ams, G.D. Marshall, M.J. Withford, Study of the influence of femtosecond laser polarisation on direct writing of waveguides. Opt. Express **14**(26), 13 158–13 163 (2006)
36. A. Politi, M.J. Cryan, J.G. Rarity, S. Yu, J.L. O'Brien, Silica-on-silicon waveguide quantum circuits. Science **320**(5876), 646–649 (2008)
37. A. Politi, J.C.F. Matthews, M.G. Thompson, J.L. O'Brien, Integrated quantum photonics. IEEE J. Sel. Top. Quantum Electron. **15**, 1673–1684 (2009)

Chapter 3
The Hong–Ou–Mandel Effect in a Waveguide Directional Coupler

3.1 Introduction

Two indistinguishable photons incident on a non-unit reflectivity beamsplitter exhibits non-classical behaviour. First experimentally measured in 1987 by Hong, Ou and Mandel [2], this behaviour lies at the heart of many linear optical quantum phenomena and is a useful benchmarking tool for the quality of photon sources and of optical circuitry.

This chapter serves as an introduction to quantum interference of degenerate photons, as well as creation operator and matrix notation that will be applied throughout this thesis to model quantum interference in more complex circuitry. This chapter also reports the first measurement of quantum interference in directly written waveguide structures; we report two photon Hong–Ou–Mandel interference and a three-photon generalisation, all implemented with directional couplers.

3.2 Classical Particles on a Beamsplitter

Suppose we launch a single particle on a beamsplitter—a semi-reflective surface—with reflectivity η and transmission $1 - \eta$. Then for either input of the beamsplitter we have two possible outcomes for detection: that the single particle is either reflected or transmitted, as shown in Fig. 3.1.

Classical intuition dictates two distinguishable particles launched into two separate input ports of the same beamsplitter have four possible detection outcomes, as shown in Fig. 3.2. This arises from probability theory that two independent events occur with probability equal to the product of the probabilities of the two events occurring individually $P(A \cap B) = P(A)P(B)$.

Some of the results reported in this chapter and an abbreviated discussion thereof were published as Refs. [1, 5].

J. C. F. Matthews, *Multi-Photon Quantum Information Science and Technology in Integrated Optics*, Springer Theses, DOI: 10.1007/978-3-642-32870-1_3,
© Springer-Verlag Berlin Heidelberg 2013

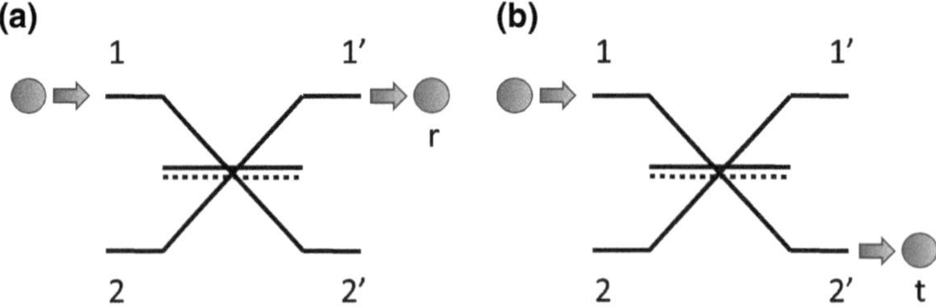

Fig. 3.1 A single particle incident to one input mode of a η reflectivity beamsplitter has two possible detection outcomes: **a** reflection (r) with η detection rate; **b** transmission (t) with $1 - \eta$ detection rate

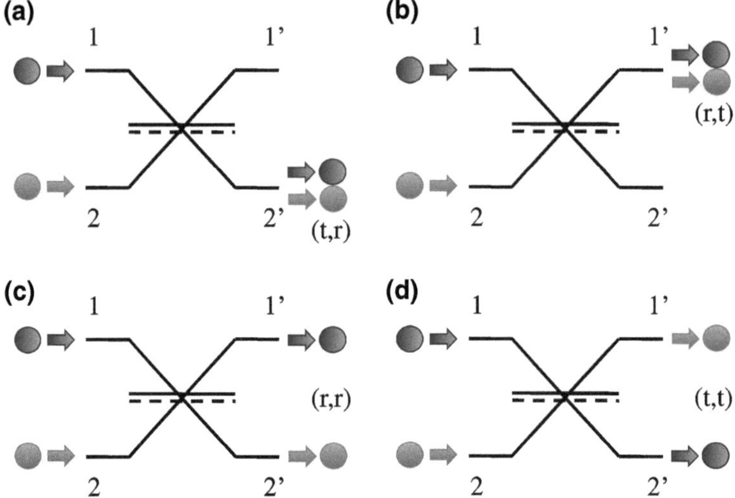

Fig. 3.2 Two classical particles (distinguished in the figure by *red* and *blue* colour) incident simultaneously to the two different input modes of a η^2 reflectivity beamsplitter have four possible detection outcomes: **a** the *red* particle is transmitted and the *blue* reflected (t,r) with detection rate $(1 - \eta)\eta$; **b** the *red* particle is reflected and the *blue* transmitted (r,t) with detection rate $\eta(1 - \eta)$; **c** both particles are reflected (r,r) with detection rate η^2; **d** both particles are transmitted (t,t) with detection rate $(1 - \eta)^2$ (Color figure online)

3.3 Two Identical Photons on a Beamsplitter

Suppose now that a pair of photons are created identical with respect to every degree of freedom (e.g. arrival time, spatial mode, polarisation and spectral properties). Launching the two photons on each of the two separate ports of a beamsplitter results in different detection outcomes to the classical case described above. Quantum mechanics dictates that the outcomes of both photons transmitted (t,t) and both photons reflected (r,r) are indistinguishable, resulting in destructive interference of the complex probability amplitudes for each outcome.

The unitary operation of a two mode beamsplitter with arbitrary reflectivity η^2 is modelled by the matrix

$$U_{BS} = \begin{pmatrix} \sqrt{\eta} & i\sqrt{1-\eta} \\ i\sqrt{1-\eta} & \sqrt{\eta} \end{pmatrix} \tag{3.1}$$

Photons created in mode j are modelled with creation operator $a_j^\dagger$ and allows description of the beam splitter as a mode transformation for each photon according to

$$a_1^\dagger |0\rangle \xrightarrow{U_{BS}} (\sqrt{\eta}a_{1'}^\dagger + i\sqrt{1-\eta}a_{2'}^\dagger) |0\rangle \tag{3.2}$$

$$a_2^\dagger |0\rangle \xrightarrow{U_{BS}} (i\sqrt{1-\eta}a_{1'}^\dagger + \sqrt{\eta}a_{2'}^\dagger) |0\rangle \tag{3.3}$$

The state of two photons launched into U_{BS} is therefore modelled [3] by

$$a_1^\dagger a_2^\dagger |0\rangle \xrightarrow{U_{BS}} (\sqrt{\eta}a_{1'}^\dagger + i\sqrt{1-\eta}a_{2'}^\dagger)(i\sqrt{1-\eta}a_{1'}^\dagger + \sqrt{\eta}a_{2'}^\dagger) |0\rangle \tag{3.4}$$

$$= (i\sqrt{\eta(1-\eta)}a_{1'}^{\dagger\,2} + \eta a_{1'}^\dagger a_{2'}^\dagger - (1-\eta)a_{2'}^\dagger a_{1'}^\dagger + i\sqrt{\eta(1-\eta)}a_{2'}^{\dagger\,2})|0\rangle \tag{3.5}$$

which is simplified by applying the Bose–Einstein commutation relations

$$[a_j^\dagger, a_k^\dagger] = 0, \ [a_j, a_k] = 0, \ [a_j, a_k^\dagger] = \delta_{j,k} \tag{3.6}$$

yielding the output state of the beamsplitter according to

$$(i\sqrt{\eta(1-\eta)}a_{1'}^{\dagger\,2} + (2\eta - 1)a_{1'}^\dagger a_{2'}^\dagger + i\sqrt{\eta(1-\eta)}a_{2'}^{\dagger\,2}) |0\rangle \tag{3.7}$$

The probability for detecting one photon at each of the outputs $1'$ and $2'$ is therefore given by

$$P^Q(1', 2') = (2\eta - 1)^2 \tag{3.8}$$

For $\eta = 0.5$, the description predicts probability $P^Q(1', 2') = 0$ of detecting one photon at each output of U_{BS}, corresponding to detecting the state $a_{1'}^\dagger a_{2'}^\dagger |0\rangle$.

To compare this to the 'classical' case of two distinguishable particles, we introduce an operator $b_j^\dagger$ acting on mode j that represents creation of a photon distinguishable (orthogonal) to the photon created by $a_j^\dagger$. It follows the action of U_{BS} on the two distinguishable particles $a_1^\dagger b_2^\dagger$, for example, is given by

$$a_1^\dagger b_2^\dagger |0\rangle \xrightarrow{U_{BS}} (i\sqrt{\eta(1-\eta)}a_{1'}^\dagger b_{2'}^\dagger + \eta a_{1'}^\dagger b_{2'}^\dagger - (1-\eta)a_{2'}^\dagger b_{1'}^\dagger + i\sqrt{\eta(1-\eta)}a_{2'}^\dagger a_{2'}^\dagger) |0\rangle \tag{3.9}$$

Since the states $a_j^\dagger |0\rangle$ and $b_j^\dagger |0\rangle$ are orthogonal, it follows that the $a_{1'}^\dagger b_{2'}^\dagger |0\rangle$ and $a_{2'}^\dagger b_{1'}^\dagger |0\rangle$ are also orthogonal. Therefore the detection event of one photon emerging at each output of U_{BS} is given by the sum of the probabilities for detecting each state $a_{1'}^\dagger b_{2'}^\dagger |0\rangle$ and $a_{2'}^\dagger b_{1'}^\dagger |0\rangle$:

$$P^C(1', 2') = \eta^2 + (1 - \eta)^2 \tag{3.10}$$

For $\eta = 0.5$, we arrive at a probability of $P^C(1', 2') = 0.5$.

Comparison between experimentally measured values of P^C and P^Q—proportional to a total number of detection events for each regime—determines the degree of quantum interference in a particular optical network. This is characterised by a visibility quantity V given by

$$V = \frac{P^C - P^Q}{P^C} \tag{3.11}$$

where P^C and P^Q are a total number of recorded detection events for the two regimes. Experimentally measured interference visibility will be bounded by an ideal visibility V_{ideal}

$$|V| < |V_{ideal}| \tag{3.12}$$

due to experimental error in making the photon pairs completely identical; V_{ideal} will be dependent upon properties of the optical network which in the case of a beam splitter is the reflectivity η.

In the experiment of Hong, Ou and Mandel [2], degenerate photon pairs incident on a beamsplitter are given a relative arrival time delay to outside of the coherence length of each photon for controlled distinguishability and to enable measurement of $P^C(1', 2')$. This delay is then tuned through zero delay for maximum overlap in order to measure $P^Q(1', 2')$. For ideal photons in the Hong–Ou–Mandel experiment, the dependence of V_{ideal} upon the reflectivity η is given by combining Eqs. (3.8, 3.10, 3.11)

$$V_{ideal} = \frac{2\eta(1 - \eta)}{1 - 2\eta + 2\eta^2} \tag{3.13}$$

In the presence of loss ε_1, ε_2, $\varepsilon_{1'}$, $\varepsilon_{2'}$ at the input and output ports of the beamsplitter, measuring reflectivity can be measured by performing a pair of experiments (Fig. 3.3) that are equivalent to four equations of four unknowns which can then be solved simultaneously. Assuming we have a fixed light intensity M input into port 1, then from input losses ε_1, ε_2 and output losses $\varepsilon_{1'}$, $\varepsilon_{2'}$ the detection scheme in Fig. 3.3 will measure two intensities

$$N_{1,1'} = M\varepsilon_1\varepsilon_{1'}\eta, \; N_{1,2'} = M\varepsilon_1\varepsilon_{2'}(1 - \eta) \tag{3.14}$$

Similarly, inputting M into port 2 yields the intensities

$$N_{2,1'} = M\varepsilon_2\varepsilon_{1'}(1 - \eta), \; N_{2,2'} = M\varepsilon_2\varepsilon_{2'}\eta \tag{3.15}$$

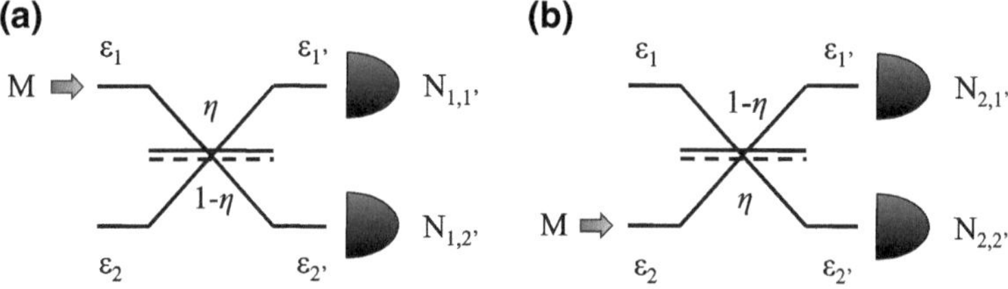

Fig. 3.3 Beamsplitter reflectivity η is measured, independent of input and output losses ε_1, ε_2, $\varepsilon_{1'}$, $\varepsilon_{2'}$, by measuring the classical intensity outputs $N_{1,1'}$, $N_{1,2'}$, $N_{2,1'}$, $N_{2,2'}$, for a fixed input intensity M

Combining Eqs. (3.14 and 3.15) yields an equation quadratic in η with solution

$$\eta = \frac{1 - \sqrt{\dfrac{N_{2,1'}N_{1,2'}}{N_{1,1'}N_{2,2'}}}}{1 - \dfrac{N_{2,1'}N_{1,2'}}{N_{1,1'}N_{2,2'}}} \tag{3.16}$$

This quantity can then be used to compute V_{ideal}.

Imperfections in the waveguide such as polarisation or spatial mode scrambling will degrade the degeneracy of the photon pairs and reduce the measured V below V_{ideal}. The relative visibility $V_{rel} \equiv V/V_{ideal}$ is therefore a measure of the fidelity of operation of a directional coupler and quantifies performance for photonic quantum circuits.

Temporal delay is not the only method for characterising quantum interference. Varied distinguishability between the two photons via any parameter is equally valid, provided it can be sufficiently manipulated. Degrees of distinguishability that cannot be controlled degrades quantum interference visibility, for example polarisation or spatial mode scrambling in waveguide for example. The ability to model distinguishability and therefore quantify the degree of photon degeneracy degradation is also useful and which can be achieved with a single parameter [4]. For example we can model distinguishability by considering an input $a_1^\dagger(\cos(\theta)a_2^\dagger + \sin(\theta)b_2^\dagger)|0\rangle$. In this way we can continuously tune between two identical photons $a_1^\dagger a_2^\dagger |0\rangle$ and two completely distinguishable photons $a_1^\dagger b_2^\dagger |0\rangle$. This formalism maps directly onto rotating relative polarisation. Proceeding in this manner, we obtain

$$a_1^\dagger(\cos(\theta)a_2^\dagger + \sin(\theta)b_2^\dagger)|0\rangle \xrightarrow{U_{BS}} \left[\cos(\theta)\left\{i\sqrt{\eta(1-\eta)}a_{1'}^{\dagger\,2} + \eta a_{1'}^\dagger a_{2'}^\dagger\right.\right.$$
$$- (1-\eta)a_{2'}^\dagger a_{1'}^\dagger + i\sqrt{\eta(1-\eta)}a_{2'}^{\dagger\,2}\Big\}$$
$$+ \sin(\theta)\Big\{i\sqrt{\eta(1-\eta)}a_{1'}^\dagger b_{2'}^\dagger + \eta a_{1'}^\dagger b_{2'}^\dagger$$
$$\left.\left.- (1-\eta)a_{2'}^\dagger b_{1'}^\dagger + i\sqrt{\eta(1-\eta)}a_{2'}^\dagger a_{2'}^\dagger\right\}\right]|0\rangle$$
$$\tag{3.17}$$

Again, since $a_j^\dagger |0\rangle$ and $b_j^\dagger |0\rangle$ are orthogonal, it follows that the probability to simultaneously detect one photon at each of the outputs $1'$ and $2'$ is given by

$$P^Q(1', 2') = \cos^2(\theta)(2\eta - 1)^2 + \sin^2(\theta)(\eta^2 + (1 - \eta)^2) \tag{3.18}$$

Combining this with Eqs. (3.10, 3.11) yields quantum interference visibility dependent upon η and a mode-mismatch parameter θ

$$\begin{aligned}
V &= \frac{(1 - \sin^2(\theta))(\eta^2 + (1 - \eta)^2) - \cos^2(\theta)(2\eta - 1)^2}{\eta^2 + (1 - \eta)^2} \\
&= \cos^2(\theta)\frac{2\eta(1 - \eta)}{1 - 2\eta + 2\eta^2} = \cos^2(\theta)V_{ideal}
\end{aligned} \tag{3.19}$$

3.4 Two Photon Quantum Interference in Directional Couplers

All linear optics can be modelled exactly using finite unitary matrices, allowing different architecture to have equivalent matrix representations. In this way, a waveguide directional coupler is represented by the matrix U_{BS} given in Eq. (3.1) and is equivalent to a bulk optical beamsplitter realised using a semi-reflective surface, for example. The Hong–Ou–Mandel experiment can therefore be tested using waveguide directional couplers.

The complete experimental setup used for measuring two photon quantum interference in a waveguide directional coupler is given schematically in Fig. 3.4. Photon pairs are generated via SPDC (see Chap. 2) at a wavelength of 804 nm collected into polarisation maintaining fibre which are connected to arrays of polarisation maintaining fibre which are spaced $250\,\mu$m apart to match the spacing of the input and output waveguides of the optical chip. The array is then butt-coupled to the optical chip using 6-axis manual nano-positioning stages (X–Y–Z translation together with pitch, yaw and roll). The output of the chip is collected using an array of single mode fibre, which is connected to fibre-coupled single photon counting modules (PerkinElmer) for coincidental counting logic.

Figure 3.4 plots the raw quantum interference data taken in a Hong–Ou–Mandel experiment using a coupler with reflectivity $\eta = 0.5128 \pm 0.0007$, computed using Eq. (3.16). The measured visibility is 0.958 ± 0.005, while the maximum achievable visibility for this reflectivity is $V_{ideal} = 0.9987 \pm 0.0001$ (using Eq. (3.13)). Seven further identical experiments are performed, for a range of different reflectivity directional couplers fabricated on two separate chips: Fig. 3.6 displays the measured visibility V as a function of the equivalent reflectivity η. The curve is a fit of Eq. (3.19) with the mode mismatch parameter $\theta = 0.218 \pm 0.001$ as a free-variable in the fit. The average relative visibility $\overline{V_{rel}} = V/V_{ideal}$ for the eight couplers is $\overline{V_{rel}} = 0.952 \pm 0.005$. We note that the non-ideal relative visibility is attributed to

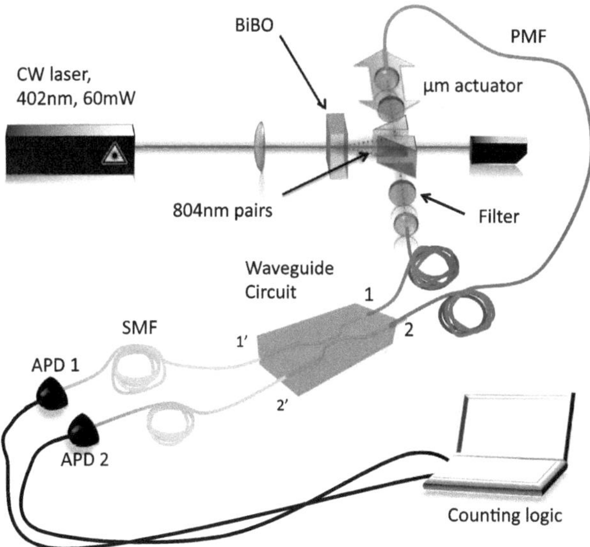

Fig. 3.4 The experimental setup for measuring Hong–Ou–Mandel quanutm interference in a directional coupler. A continuous wave 402 nm diode laser pumps a nonlinear BiBO crystal for type-I down conversion, to generate approximately single pairs of degenerate photons (804 nm, see Chap. 2). The photons are collected into polarisation maintaining fibre (PMF), which are buttcoupled to the waveguide chip using arrays of PMF. The *green arrow* indicates a micrometer actuator used to control the relative temporal delay between the arrival times of the degenerate photon pair. Figure reproduced from Ref. [5] (Colour figure online)

an imperfect experimental setup, rather than to the directional couplers; for example slight distinguishability of the photon pairs in polarisation and wavelength and the data given is not corrected for accidental coincidence counts arising from the five nano-second coincidence window in the counting electronics.

3.5 Three Photon Quantum Interference in a Directional Coupler

Hong–Ou–Mandel interference generalises in many forms, including varying the optical circuit or the number of degenerate photons used in the experiment. Here we report an example of a three-photon interference experiment, first demonstrated in bulk optics [6], which we realise in a femtosecond laser machined waveguide directional coupler (see Chap. 2) [5]. Analogous to the Hong–Ou–Mandel experiment, three photons are incident into a beamsplitter with the initial state $|2\rangle_1 |1\rangle_2$; provided the beamsplitter has reflectivity $\eta = 2/3$, destructive interference leads to an absence

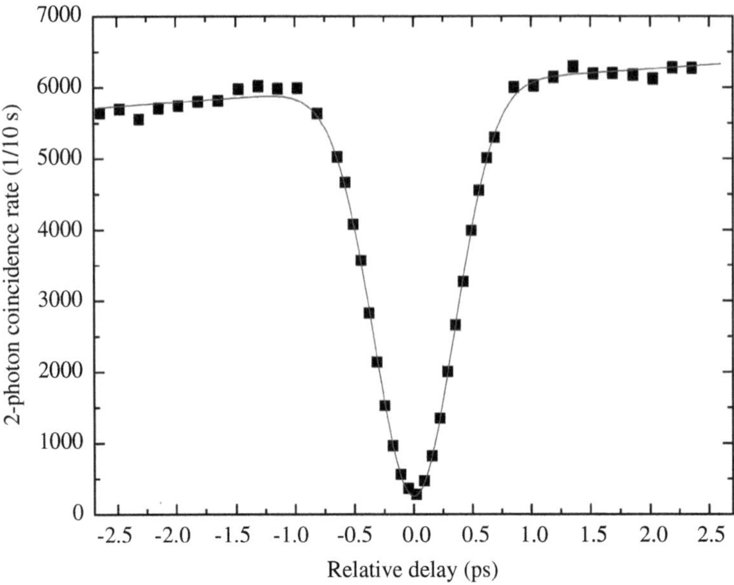

Fig. 3.5 Hong–Ou–Mandel quantum interference in a laser direct-write waveguide directional coupler. The number of coincident detections across the output of the coupler are plotted as a function of the relative temporal delay between the two input photons. *Error bars* are from assuming Poissonian statistics on the coincidence rate and are smaller than the point size. The fit is Gaussian plus a linear curve to account for decoupling of the photon pair source as the collecting fibre in the source is moved. Figure reproduced from Ref. [5]

of the state $|2\rangle_1 |1\rangle_2$ at the output of the device. This is modelled according to

$$\frac{1}{\sqrt{2!}} a_1^{\dagger 2} a_2^{\dagger} |0\rangle \xrightarrow{U_{BS}} \left(\frac{1}{\sqrt{2!}} (\eta\, a_{1'}^{\dagger} + i\sqrt{1-\eta}\, a_{2'}^{\dagger})^2 (i\sqrt{1-\eta}\, a_{1'}^{\dagger} + \eta\, a_{2'}^{\dagger}) \right) |0\rangle \tag{3.20}$$

$$= \left(\frac{i}{3}\sqrt{\frac{2}{3}} a_{1'}^{\dagger 3} + \frac{i}{\sqrt{6}} a_{1'}^{\dagger} a_{2'}^{\dagger 2} - \frac{1}{3\sqrt{3}} a_{2'}^{\dagger 3} \right) |0\rangle \tag{3.21}$$

$$= i\frac{2}{3} |3\rangle_{1'} |0\rangle_{2'} + \frac{i}{\sqrt{3}} |1\rangle_{1'} |2\rangle_{2'} - \frac{\sqrt{2}}{3} |0\rangle_{1'} |3\rangle_{2'} \tag{3.22}$$

for three indistinguishable photons with associated creation operators $a_j^{\dagger}$. In contrast, we can consider two further cases where we introduce one distinguishable photon along with two indistinguishable photons: $\frac{1}{\sqrt{2!}} a_1^{\dagger 2} c_2^{\dagger} |0\rangle$ and $a_1^{\dagger} b_1^{\dagger} a_2^{\dagger} |0\rangle$, where $\langle 0| a_j b_j^{\dagger} |0\rangle = 0$, $\langle 0| a_j c_j^{\dagger} |0\rangle = 0$ and $\langle 0| b_j c_j^{\dagger} |0\rangle = 0$. Launching the state $\frac{1}{\sqrt{2!}} a_1^{\dagger 2} c_2^{\dagger} |0\rangle$ is modelled according to

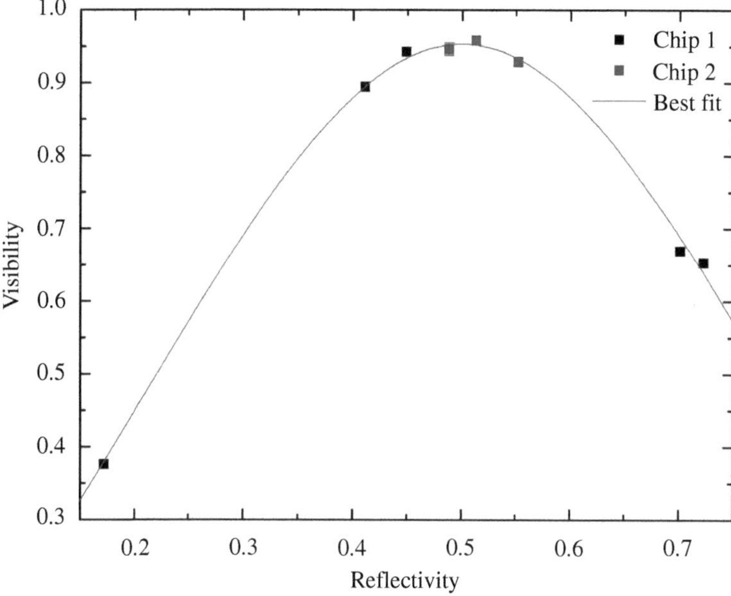

Fig. 3.6 Quantum interference visibility for Hong–Ou–Mandel experiments performed using a range of differing reflectivity η directional couplers fabricated in two different monolithic chips. Each data-point represents one experiment which is fitted as in Fig. 3.5 and *error bars* are computed using 95 % confidence on the fitting parameters. *Error bars* are smaller than the data points. Figure reproduced from Ref. [5]

$$\frac{1}{\sqrt{2!}}a_1^{\dagger\,2}c_2^{\dagger}\,|0\rangle \xrightarrow{U_{BS}} \left(-\frac{2}{3\sqrt{3}}\underline{a_{1'}^{\dagger}a_{2'}^{\dagger}c_{1'}^{\dagger}} + \frac{2}{3\sqrt{3}}a_{1'}^{\dagger\,2}c_{2'}^{\dagger} - \frac{1}{3\sqrt{3}}a_{2'}^{\dagger\,2}c_{2'}^{\dagger} \right.$$

$$\left. + \frac{i}{3}\sqrt{\frac{2}{3}}a_{1'}^{\dagger\,2}c_{1'}^{\dagger} - \frac{i}{3\sqrt{6}}a_{2'}^{\dagger\,2}c_{1'}^{\dagger} + \frac{2i}{3}\sqrt{\frac{2}{3}}a_{1'}^{\dagger}a_{2'}^{\dagger}c_{2'}^{\dagger} \right)\,|0\rangle$$

$$(3.23)$$

where we have underlined the terms that give rise to detecting two photons at output 1 and one photon at output 2. The probability for such a detection event is $4/27+8/27 = 4/9$. Launching the state $a_1^{\dagger}b_1^{\dagger}a_2^{\dagger}\,|0\rangle$ is modelled according to

$$a_1^{\dagger}b_1^{\dagger}a_2^{\dagger}\,|0\rangle \xrightarrow{U_{BS}} \left(\frac{1}{3}\sqrt{\frac{2}{3}}\underline{a_{1'}^{\dagger}b_{1'}^{\dagger}a_{2'}^{\dagger}} - \frac{1}{3}\sqrt{\frac{2}{3}}a_{1'}^{\dagger\,2}b_{2'}^{\dagger} - \frac{1}{3}\sqrt{\frac{2}{3}}a_{2'}^{\dagger\,2}b_{2'}^{\dagger} \right. \tag{3.24}$$

$$\left. + \frac{2i}{3\sqrt{3}}a_{1'}^{\dagger\,2}b_{1'}^{\dagger} + \frac{2i}{3\sqrt{3}}b_{1'}^{\dagger}a_{2'}^{\dagger\,2} + \frac{i}{3\sqrt{3}}a_{1'}^{\dagger}a_{2'}^{\dagger}b_{2'}^{\dagger} \right)\,|0\rangle \tag{3.25}$$

The probability to detect two photons at output 1 and one photon at output 2 for this case is $2/27 + 4/27 = 2/9$.

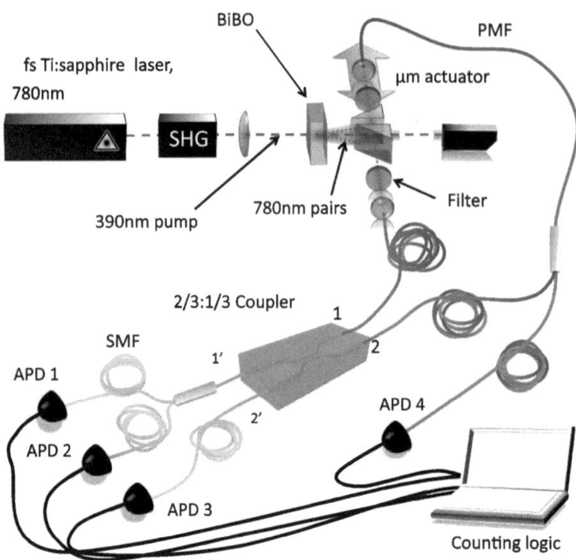

Fig. 3.7 Experimental setup for measuring three-photon quantum interference in a waveguide directional coupler. The multi-photon state is produced via pulsed spontaneous parametric down conversion (see Chap. 2). The photons are launched into and collected from the chip using butt-coupled arrays of optical fibre. Figure reproduced from Ref. [5]

The fourth case we consider is to make all three input photons as distinguishable, modelled according to

$$a_1^\dagger b_1^\dagger c_2^\dagger |0\rangle \xrightarrow{U_{BS}} \left(-\frac{1}{3}\sqrt{\frac{2}{3}} \underline{b_{1'}^\dagger c_{1'}^\dagger a_{2'}^\dagger} - \frac{1}{3}\sqrt{\frac{2}{3}} \underline{a_{1'}^\dagger c_{1'}^\dagger b_{2'}^\dagger} + \frac{2}{3}\sqrt{\frac{2}{3}} \underline{a_{1'}^\dagger b_{1'}^\dagger c_{2'}^\dagger} - \frac{1}{3}\sqrt{\frac{2}{3}} a_{2'}^\dagger b_{2'}^\dagger c_{2'}^\dagger \right.$$

$$\left. + \frac{2i}{3\sqrt{3}} a_{1'}^\dagger b_{1'}^\dagger c_{1'}^\dagger - \frac{i}{3\sqrt{3}} c_{1'}^\dagger a_{2'}^\dagger b_{2'}^\dagger + \frac{2i}{3\sqrt{3}} b_{1'}^\dagger a_{2'}^\dagger c_{2'}^\dagger + \frac{2i}{3\sqrt{3}} a_{1'}^\dagger b_{2'}^\dagger c_{2'}^\dagger \right) |0\rangle$$

$$(3.26)$$

For this case, the probability to detect two photons at output 1 and one photon at output 2 is $2/27 + 2/27 + 8/27 = 4/9$. From this model, we can conclude that a Hong–Ou–Mandel style dip in detecting two photons at output 1 and one photon at output 2 can be observed, by launching three indistinguishable photons in the state $\frac{1}{\sqrt{2!}} a_1^{\dagger 2} a_2^\dagger |0\rangle$ into a $\eta = 2/3$ reflectivity beamsplitter and delaying, for example, the photon input into port 2, in order to input the state $\frac{1}{\sqrt{2!}} a_1^{\dagger 2} b_2^\dagger |0\rangle$. This experiment has been previously reported using bulk optics [6].

Figure 3.7 displays the experimental setup for launching the states $\frac{1}{\sqrt{2!}} a_1^{\dagger 2} a_2^\dagger |0\rangle$ and $\frac{1}{\sqrt{2!}} a_1^{\dagger 2} b_2^\dagger |0\rangle$ into an $\eta = 2/3$ reflectivity waveguide directional coupler.

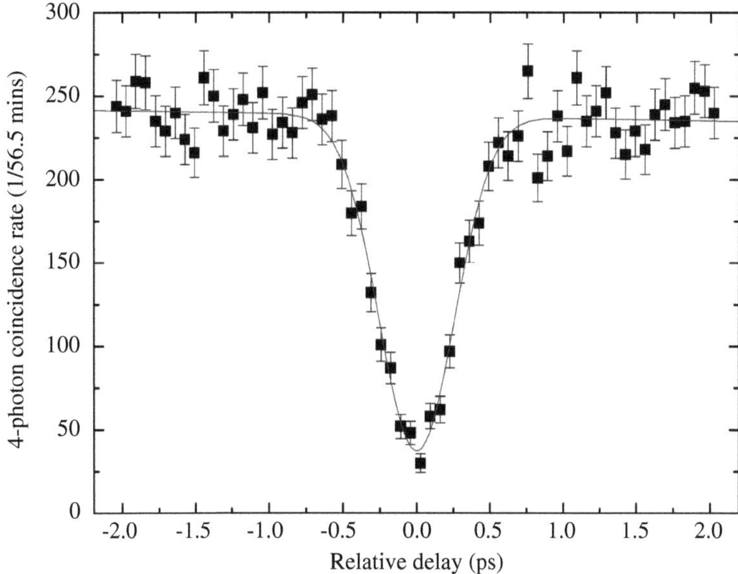

Fig. 3.8 Generalised Hong–Ou–Mandel interference of three photons measured in an reflectivity $\eta = 0.659$ directional coupler. The four-photon detection events plotted on the vertical axis correspond to detecting two photons at output $1'$ and one photon at output $2'$ of the directional coupler. The horizontal axis corresponds to relative optical delay between inputs 1 and 2. Figure reproduced from Ref. [5]

The three photon state is post-selected from the four photon term $|2\rangle\,|2\rangle$ in the four photon down conversion state, output from a two-mode, pulsed Ti:sapphire multi-photon spontaneous parametric down conversion (SPDC) source (see Chap. 2). One of the collected modes—coupled into polarisation maintaining fibre (PMF)—is split non-deterministically using a PMF splitter of which one output is connected directly to one avalanche photodiode (APD 4) single photon counting module. The remaining fibre from the splitter is connected to the input of a $250\,\mu$m spaced polarisation maintaining fibre array (not shown). The second collecting fibre from the SPDC source is connected to the neighbouring fibre in the array; the array is then butt-coupled to the inputs of an $\eta = 2/3$ directional coupler. The two optical fibres guiding photons output from the SPDC, into the waveguide chip are coarsely matched in length with a tolerance of $\sim$5 mm; the optical delay between photons input into modes 1 and 2 is then tuned to zero using a micrometer actuator—translating one collecting fibre along the direction of incoming photons—and a reduced visibility, two photon Hong–Ou–Mandel experiment. The four-photon coincidence event is then recorded with APDs 1–4 as shown in Fig. 3.7, using non-deterministic number resolving detection via an optical fibre splitter and APDs 1 and 2, to detect two photons in output $1'$.

Figure 3.8 shows the generalized HOM dip observed in a $\eta = 0.659$ reflectivity direct-write coupler. The visibility of this dip is $V = 0.84 \pm 0.03$ (compared with $V = 0.78 \pm 0.05$ in a previous bulk optics implementation [6]). The non-unit visibility is attributed mainly to a degree of temporal distinguishability of the two photons produced in the source in the $|2\rangle_1$ state.

3.6 Discussion

We have demonstrated quantum interference of two and three photon states in femtosecond laser machined waveguide directional couplers. Together with previous results [7] reporting $V = 94.8\,\%$ visibility Hong–Ou–Mandel interference in similar structures implemented using standard lithographic techniques (see Chap. 2)—subsequently improved to unit visibility [8]—waveguide directional couplers satisfy one of the key requirements for constructing linear optical experiments to test quantum mechanics and to realise proof-of principle quantum technologies. We have reported controlled quantum interference over a range of devices by varying coupler reflectivity and we have reported an improved interference visibility in a particular example of multi-photon quantum interference. The following chapters investigate further photonic quantum interference in more complex waveguide structures. While the remainder of this thesis is concerned with experiments employing lithographically fabricated waveguide circuits, femto-second laser machined waveguide structures do overcome certain limitations of standard lithographic approaches, opening the way to three dimensional structures and the potential for rapid prototyping of individual circuits.

References

1. G.D. Marshall, P. Dekker, M. Ams, J.A. Piper, M.J. Withford, Directly written monolithic waveguide laser incorporating a distributed feedback waveguide-bragg grating. Opt. Lett. **33**(9), 956–958 (2008)
2. C.K. Hong, Z.Y. Ou, L. Mandel, Measurement of subpicosecond time intervals between two photons by interference. Phys. Rev. Lett. **59**(18), 2044–2046 (1987)
3. H.A. Bachor, T.C. Ralph, *A Guide to Experiments in Quantum Optics*, 2nd edn. (Wiley, New York, 2004)
4. P.P. Rohde, G.J. Pryde, J.L. O'Brien, T.C. Ralph, Quantum-gate characterization in an extended hilbert space. Phys. Rev. A **72**(3), 032306 (2005)
5. G.D. Marshall, A. Politi, J.C.F. Matthews, P. Dekker, M. Ams, M.J. Withford, J.L. O'Brien, Laser written waveguide photonic quantum circuits. Opt. Express **17**(15), 12546–12554 (2009)
6. K. Sanaka, K.J. Resch, A. Zeilinger, Filtering out photonic fock states. Phys. Rev. Lett. **96**(8), 083601 (2006)
7. A. Politi, M.J. Cryan, J.G. Rarity, S. Yu, J.L. O'Brien, Silica-on-silicon waveguide quantum circuits. Science **320**(5876), 646–649 (2008)
8. A. Laing, A. Peruzzo, A. Politi, M. Rodas Verde, M. Halder, T.C. Ralph, M.G. Thompson, J.L. O'Brien, High-fidelity operation of quantum photonic circuits. Appl. Phys. Lett. **97**, 211109 (2010)

Chapter 4
Multi Directional-Coupler Circuit for Quantum Logic

4.1 Introduction

The previous chapter discussed quantum interference using waveguide directional couplers as an integrated version of a beam splitter. The beam splitter is one of the building blocks of any quantum optical technology, including linear optical quantum computing. In this chapter, we discuss the use of multiple directional couplers to create quantum optical logic. Specifically we demonstrate the use of two such coupler based logic gates in a single Silica-on-Silicon waveguide chip (see Sect. 2.4.2) to realise a proof of principle demonstration of a subroutine of Shor's quantum factoring algorithm to factorize 15.

Peter Shor's quantum factoring algorithm [2] determines the prime factors of a large number exponentially faster than any other known method—a computing task that lies at the heart of modern information security. Shor's full algorithm requires a full-scale quantum computer to harnesses the 'massive parallelism' afforded by quantum superposition and entanglement of quantum information.

Despite progress towards this goal, proof-of-principle (and non-scalable) *compiled demonstrations* of Shor's algorithm have so far only been possible with liquid-state nuclear magnetic resonance [3] and bulk optical implementations of simplified logic gates [4, 5], owing to the need for several logic gates operating on several qubits, even for small-scale compiled versions. To date, these approaches cannot be scaled to a large number of qubits due to the purity, size and stability limitations of these systems. Furthermore, before attempting to perform Shor's algorithm in a fault tolerant regime, the full algorithm for factorizing 15 requires ∼4000 quantum gates and 21 qubits [5].

In the short term, implementing compiled versions of algorithms serve as a benchmark of the level of complexity achievable by quantum platforms due to the complexity requirements of several operational quantum logic gates and multiple qubits.

Some of the results reported in this chapter and an abbreviated discussion thereof were published as Ref. [1].

J. C. F. Matthews, *Multi-Photon Quantum Information Science and Technology in Integrated Optics*, Springer Theses, DOI: 10.1007/978-3-642-32870-1_4,
© Springer-Verlag Berlin Heidelberg 2013

They also allow the community to test aspects of a given quantum algorithm and can be considered to be tests of a subroutine of the full algorithm using the successful operation of the required multiple logic gates and qubits.

4.2 A Compiled Version of Shor's Algorithm

Factorisation is simple to define: given a product of two unknown prime numbers $N = pq$, determine the constituent factors p and q. Solving factorisation equates to finding the parameter r known as the order of a given coprime a of N. Euler's theorem dictates that for all coprime integers a and N (no common factor), there exists r such that

$$a^r \equiv 1 \bmod N \text{ and } 1 \le r < N \tag{4.1}$$

Provided r is even, then it follows that

$$a^r - 1 = \left(a^{r/2} - 1\right)\left(a^{r/2} + 1\right) \equiv 0 \bmod N \tag{4.2}$$

which implies N divides the product $\left(a^{r/2} - 1\right)\left(a^{r/2} + 1\right)$. Since r is the order of a modulo N, and provided $a^{r/2} \ne -1 \bmod N$, then it follows that the factors of N must each divide $\left(a^{r/2} - 1\right)$ and $\left(a^{r/2} + 1\right)$. This implies the factors of N are given by the greatest common divisors of N and $\left(a^{r/2} \pm 1\right)$ which can be computed efficiently on a classical computer using Euclid's algorithm. The most challenging part of Shor's algorithm, requiring the power of quantum computation, is to find the order r of some randomly chosen coprime a of N.

There is no known classical algorithm that can achieve this in time or resources polynomial in N. However, the order finding routine of Shor's quantum algorithm uses entanglement and superposition across two registers of $s = 2\log_2 N$ qubits (known as the argument register x_i and the function register f_i) to compute the modular exponential function (MEF) $a^z \bmod N$ in a polynomial number of resources and time. The routine initialises the first register as a superposition of all the states in the s–qubit computational basis, and applies a so-called MEF unitary transformation that evolves the state to

$$\frac{1}{\sqrt{2^s}} \sum_{x=0}^{2^s-1} |x\rangle \left|a^x \bmod N\right\rangle \tag{4.3}$$

This is a highly entangled state. At this stage a quantum Fourier transform (QFT) is applied on the first register to obtain

Fig. 4.1 The quantum circuit for a compiled version of Shor's algorithm. Figure adapted from Ref. [1]

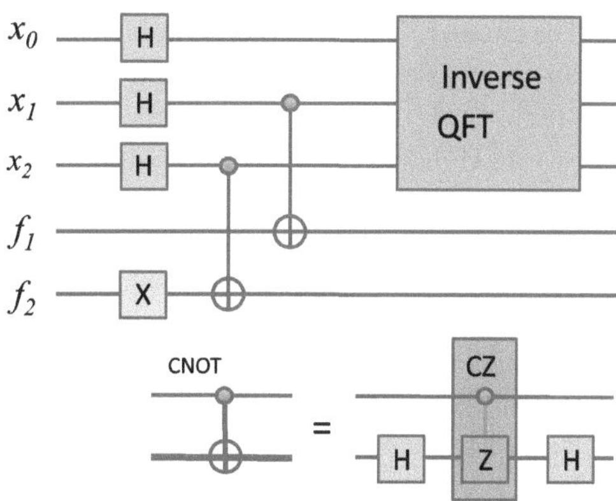

$$\frac{1}{\sqrt{2^s}} \sum_{x=0}^{2^s-1} \sum_{y=0}^{2^s-1} e^{2i\pi xy} |y\rangle \, |a^x \bmod N\rangle \qquad (4.4)$$

From this state, interference between the terms leads to amplitude peaks for $y = c2^s/r$, $c \in \mathbb{Z}$ and therefore the order r is found with high probability [6].

It is important to note that the entire routine fails for two conditions: (i) r is not even or (ii) $a^{r/2} = -1 \bmod N$. In either case, factorisation has the benefit that verification is simple since multiplication of two proposed factors p and q is efficient to do on a classical computer. Furthermore, a key element of the algorithm is that the initial co-prime a is chosen at random. Provided this is the case, then it is shown that the whole routine has a probability of $P > 1/2$ of success. Therefore a linear number n of repetitions of the routine increases the probability of success towards unity according to $P > 1 - 1/2^n$ [2, 6].

Our demonstration factorizes 15 using a compiled version of the order finding routine in Shor's algorithm. The method of compiling Shor's algorithm arises from the fact that each coprime a of N does not require the full circuit for the whole algorithm [7]. The circuit given in Fig. 4.1 was previously implemented in a bulk optical architecture [5] and is designed specifically factorize 15 by finding the order of the coprime $a = 2$, which does lead to the correct calculation of the factors of 15. This compilation reduces the required number of function qubits (f_i) by evaluating $\log_a[a^x \bmod N]$ in place of $a^x \bmod N$; reducing the number of function qubits from $\log_a[N]$ to $\log_2[\log_a[N]]$. Note that the functions $\log_a[a^x \bmod N]$ and $a^x \bmod N$ have the same period. The main components of this five qubit circuit are two controlled-NOT (CNOT) gates (discussed in Sect. 4.3), single qubit operations and an inverse quantum Fourier transform (QFT). The circuit is further simplified by the fact that the inverse QFT shown in Fig. 4.1 is not needed in the circuit for our case (nor for any $r = 2^l$, $l \in \mathbb{N}$), and can be performed by classical post processing, as shown in the supplemental material of Ref. [5].

The circuit [5] operates on five qubits, one of which, x_0, is effectively redundant as it remains in a separable state throughout and is only used in the classical implementation of the QFT. Inputting the state $|\psi_{in}\rangle = |0\rangle_{x1} |0\rangle_{x2} |0\rangle_{f1} |1\rangle_{f2}$ produces a product of two maximally entangled Bell pairs: $\frac{1}{2}(|0\rangle_{x1} |0\rangle_{f1} + |1\rangle_{x1} |1\rangle_{f1})(|0\rangle_{x2} |1\rangle_{f2} + |1\rangle_{x2} |0\rangle_{f2})$. Tracing out the f_i qubits allows the outputs of the quantum computation to be obtained by measuring the x_i qubits in the computational basis, which combined with the qubit x_0 in the zero state, yields the 3-bit output: $x_2, x_1, x_0 = 000, 010, 100$, or 110. These 3-bit numbers correspond to the values 0, 2, 4 and 6, respectively. The first number is an expected failure inherent to Shor's algorithm, while the third yields the trivial factors 1 and 15. The second and fourth outcomes allow the calculation of the order $r = 4$ of $a = 2$, which via Euclid's classical algorithm—efficient on a classical computer—yields the correct prime factors 3 and 5. Therefore a single implementation of the routine, with deterministic gates, has an overall success rate of 1/2.

4.3 Linear Optical CNOT Gate

The ability to entangle qubits is fundamental for quantum computing, quantum information processing and quantum mechanical technologies in general. Any candidate platform for quantum information science must provide a means for generating entanglement and in the circuit model of quantum computation, the two-qubit controlled NOT (CNOT) gate is the basis for achieving this.

Together with arbitrary single qubit rotations, the CNOT gate forms the universal set of gates for quantum computation and gains its importance from its ability to manipulate the state of two qubits—known as the control qubit (C) and the target qubit (T)—in such a way that the final output can be entangled. For this reason the performance of the CNOT gate has become a benchmark to evaluate the capabilities of quantum platforms. In the computational basis, the operation of a CNOT gate is simple; the state of the target qubit is flipped, conditional on the logical state of the control qubit:

$$|0\rangle_C |0\rangle_T \xrightarrow{\text{CNOT}} |0\rangle_C |0\rangle \tag{4.5}$$

$$|0\rangle_C |1\rangle_T \xrightarrow{\text{CNOT}} |0\rangle_C |1\rangle \tag{4.6}$$

$$|1\rangle_C |0\rangle_T \xrightarrow{\text{CNOT}} |1\rangle_C |1\rangle \tag{4.7}$$

$$|1\rangle_C |1\rangle_T \xrightarrow{\text{CNOT}} |1\rangle_C |0\rangle \tag{4.8}$$

The ability of the CNOT gate to entangle two qubits becomes clear when a coherent superposition state is input as the target qubit, for example inputting the state $|+\rangle_C |0\rangle_T$ generates a maximally entangled Bell state

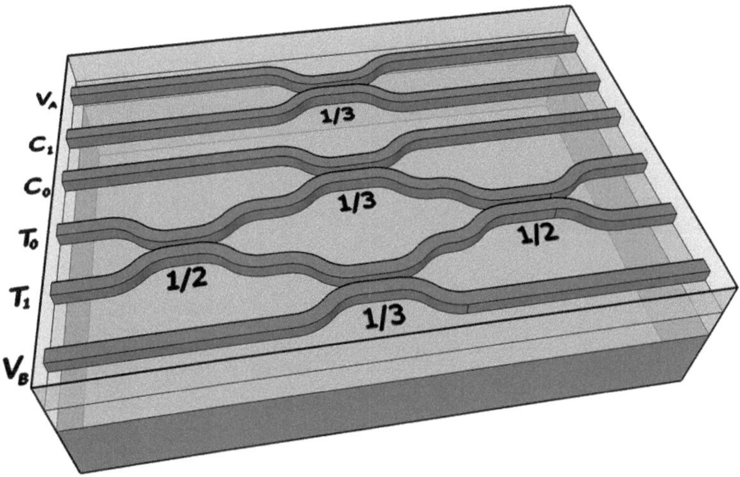

Fig. 4.2 Schematic representation of a linear optical CNOT gate, realised in waveguide using directional couplers (with labeled reflectivity) to realise the required beamsplitters [16]. Figure reproduced from Ref. [18]

$$\frac{1}{\sqrt{2}}\left(|0\rangle_C + |1\rangle_C\right)|0\rangle_T \xrightarrow{\text{CNOT}} \frac{1}{\sqrt{2}}\left(|0\rangle_C |0\rangle + |1\rangle_C |1\rangle\right) \qquad (4.9)$$

A post-selected version of the CNOT gate was proposed that consists only of linear optical components [8, 9] and has since been implemented using bulk-optical elements [10–14] using a range of methods to avoid instability of the central interferometer and applied to proof of principle demonstrations of quantum computing, for example compiled versions of Shor's algorithm [4, 5]. Such implementations typically occupy $\sim 1\,\mathrm{m}^2$ of bulk optical elements bolted down on an optical bench, potentially limiting the scalability of such an approach. This has driven miniaturised realisations using polarisation encoding in optical fibre [15] and waveguide circuits integrated into monolithic chips [16–18]. Figure 4.2 displays the original circuit proposed in Refs. [8, 9] depicted with waveguide circuitry [16–18].

The linear optical [8, 9] CNOT gate works as follows: control C and target T qubits are each encoded by a photon in two waveguides C_0, C_1 and T_0, T_1. Two $\eta = 1/2$ reflectivity beamsplitters in target waveguides (T_0 and T_1) form a balanced Mach Zehnder Interferometer to perform either the identity operation or the NOT operation on T, conditional on a phase shift in the interferometer. For C in the logical one state (photon in the waveguide C_1), the control and target photons interfere nonclassically at the central $\eta = 1/3$ reflectivity beamsplitter. For $\eta = 1/3$, the resulting state evolves via quantum interference to $|1\rangle_C |0\rangle_T \rightarrow -1/3(|1\rangle_C |0\rangle_T)$, conditional on post-selecting one photon in C_0 or C_1 and one photon in T_0 or T_1.

This causes a π phase difference between the arms of the interferometer, and so performing a NOT operation on the target qubit. Of course, due to post-selection, the presented scheme does not work with certainty: it is possible, for example, to

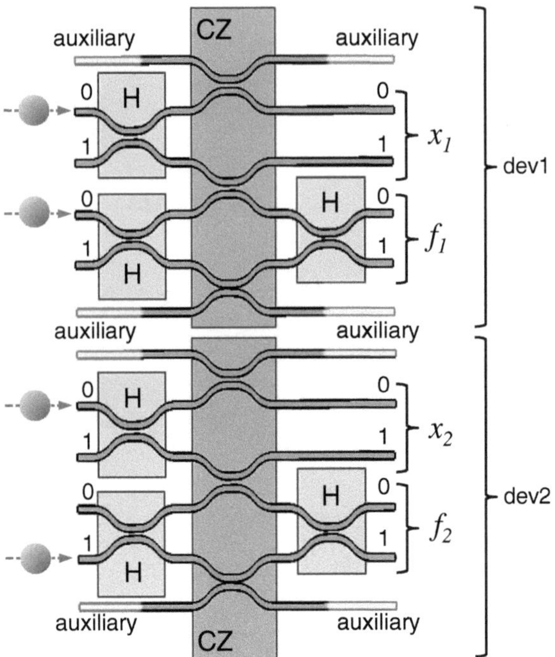

Fig. 4.3 Schematic of the waveguide circuit that implements the compiled version of Shor's algorithm. The x_n qubits carry the result of the algorithm; f_n qubits form the function register of additional qubits required for the computation. Figure reproduced from Ref. [1]

find two photons in the control or target outputs. Operation of the gate therefore relies on the ability to detect the presence of only one photon in either of the control modes and one photon in either of the target modes to herald the success of the gate. The probability of such an event is equal to 1/9 for all the possible combinations of input states, given by two further $\eta = 1/3$ reflectivity couplers that balance the overall gate success probability. This circuit has become particularly important for quantum optical platforms as a benchmark, since it relies on both quantum interference at the central 1/3 beamsplitter and classical interference in the Mach Zehnder interferometer acting on the target qubit.

4.4 Realising the Compiled Algorithm with a Waveguide Circuit

The physical implementation of a compiled version of the order finding routine for Shor's factoring algorithm to factor 15 for coprime $a = 2$ and order $r = 4$ is shown in Fig. 4.3. It consists of two non-deterministic controlled-phase (CZ) gates (each with success $P = 1/9$, conditional on post selection) and six one-qubit complex Hadamard (H) gates [19]. The computation proceeds as follows: four photons are input into the "0" or "1" waveguides to prepare the initial state

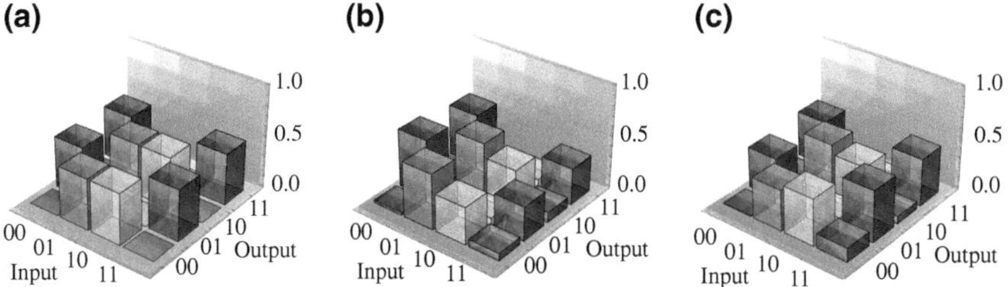

Fig. 4.4 Performance of dev1 and dev2. **a** The ideal truth table for the two devices. The measured logical truth tables of gates dev1 (**b**) and dev2 (**c**) in Fig. 4.3, used to realise a compiled version of Shor's algorithm

$|\psi_{in}\rangle = |0\rangle_{x_1}|0\rangle_{x_2}|0\rangle_{f_1}|1\rangle_{f_2}$ (this does not represent the number 15, but rather the initialization for the compiled algorithm to compute the factors of 15). The H gates, implemented by $\eta = 1/2$ reflectivity directional couplers, then prepare each qubit in a superposition of 0 and 1, such that the entire state is a superposition of all possible four-bit inputs. The core process is then performed by two independent CZ gates, each implemented by a network of three $\eta = 1/3$ reflectivity directional couplers as described in Sect. 4.3, that create a highly entangled output state [4, 5]. Measurement of the output state of qubits x_1 and x_2, and classical processing, gives the results of the computation. The network of CZ and H gates used to realise the compiled algorithm are integrated in one silica-on-silicon waveguide chip; the circuit is drawn schematically in Fig. 4.3 and consists of two copies of a single circuit (labeled dev1 and dev2).

The qubits used for the algorithm were realised with two pairs of 790 nm photons generated simultaneously via spontaneous parametric down conversion, collected into two pairs of polarisation maintaining fibre (Chap. 2). The photons were then coupled into and out of the chip with butt-coupled arrays of optical fibre and detected at the end of the integrated device using Silicon avalanche photo diode single photon counting modules.

Devices dev1 and dev2 were each measured in the computational basis using photon pairs. The quantum interference condition required for circuit operation ensured using a Hong–Ou–Mandel experiment [20]. The truth tables for dev1 and dev2 are given in Fig. 4.4 and with the ideal truth table they have fidelities of $S = 87.8 \pm 1.1\%$ and $S = 89.4 \pm 1.3\%$ respectively, where fidelity between ideal and measured distributions is given by $S = (\sum_{i,j} \sqrt{P_{i,j} P'_{i,j}})^2 / \sum_{i,j} P_{i,j} \sum_{i,j} P'_{i,j}$. We attribute the non-perfect similarity to imperfect splitting ratios in the composite directional couplers and slight distinguishability of the incident photons.

To run the quantum algorithm we input the four qubits in the logical "0" state; the measured output statistics of qubits x_1 and x_2—together with the qubit x_0 that remains in the 0 state—in the binary string $x_2 x_1 x_0$, are presented in Fig. 4.5, showing the four outcomes $y = 000 = 0$, $y = 010 = 2$, $y = 100 = 4$, $y = 110 = 6$ (expressed in both binary and decimal notation). The reversal of the outcome order

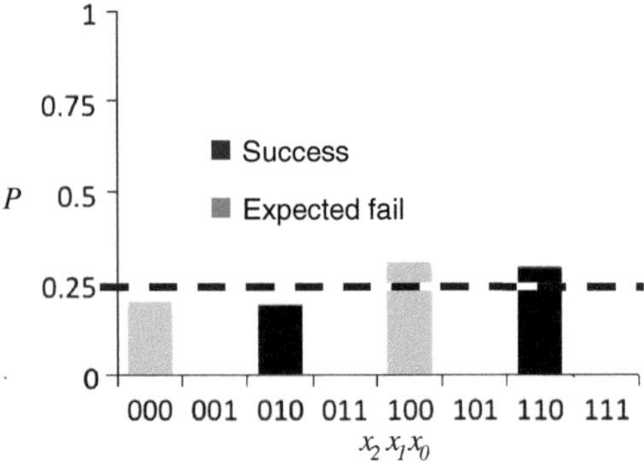

Fig. 4.5 Measured outcomes of the compiled algorithm consisting of the two expected success and two expected fails of the algorithm. The success of the demonstration is measured by the similarity distribution of the output events and the ideal distribution (*dashed line*) which in this demonstration is a uniform probability of 1/4 for outcomes corresponding to 000, 010, 100 and 111 and a probability of 0 for the remaining possible outcomes

is the only step required for the simplified inverse quantum Fourier transform [5]. Outputs $y = 010 = 2$ and $y = 110 = 6$ lead to the correct calculation for finding the order $r = 4$ of $a = 2$ for the algorithm, which then enables efficient classical computation of the factors 3 and 5; $y = 100 = 4$ gives the trivial factors (1 and 15); and $y = 000 = 0$ is an expected failure mode inherent to Shor's algorithm. Calculation of r, as detailed for example in [6], is computed from the expression $y/2^s = c/r$ where c/r, for some integer c, is a fraction in its simplest form and s is the number of qubits in the argument register ($s = 3$ in our case). The continued fractions expansion (which is efficiently computable on a classical computer) can then be used to extract r; in our case we have $y/8 = 2/8 = 1/4$ and $y/8 = 6/8 = 3/4$, in both cases yielding $r = 4$ from the denominator of the reduced fraction. The output result shows a similarity of $S = 98.9 \pm 0.8\,\%$ between the measured distribution P and the ideal distribution P (dashed line). We note that this distribution can be computed for example with classical means, however full quantum process tomography of the scheme we have demonstrated here has been reported [5], demonstrating the required presence of entanglement.

4.5 Discussion

This demonstration of a small-scale compiled version of Shor's algorithm illustrates that multiple directional coupler based circuits can be integrated together to perform simple quantum logic gate and shows great promise for quantum technologies using integrated waveguide circuits. Our results report the interferometric stability (classical interference) and quantum interference conditions required to realise quantum circuits of similar complexity. Any quantum computer is a many-particle, many-path interferometer; the capability to implement such complex interferometers in a stable and miniaturized architecture is therefore critical to the future realization of large-scale quantum algorithms.

However, the measurement of the circuit was performed in the logical computational basis alone. While quantum coherence inside identical structures has previously been reported[16, 18], there is now need of a means for reconfigurability of waveguide circuits to perform state and process tomography as was carried out in bulk optical implementation of the experiment reported here[5] and indeed other quantum technology circuits[11]. The following chapter begins to address this by testing a device equivalent to a reconfigurable beam splitter.

References

1. A. Politi, J.C.F. Matthews, J.L. O'Brien, Shor's quantum factoring algorithm on a photonic chip. Science **325**(5945), 1221 (2009)
2. P. W. Shor, Algorithms for quantum computation: discrete logarithms and factoring, in *Proceedings of 35th Annual Symposium Foundations of Computer Science*, ed. by S. Goldwasser (IEEE Computer Society, Los Alamitos, 1994), pp. 124–134
3. L.M.K. Vandersypen, M. Steffen, G. Breyta, C.S. Yannoni, M.H. Sherwood, I.L. Chuang, Experimental realization of shor's quantum factoring algorithm using nuclear magnetic resonance. Nature **414**, 883–887 (2001)
4. C.-Y. Lu, D.E. Browne, T. Yang, J.W. Pan, Demonstration of a compiled version of shor's quantum factoring algorithm using photonic qubits. Phys. Rev. Lett. **99**, 250504 (2007)
5. B.P. Lanyon, T.J. Weinhold, N.K. Langford, M. Barbieri, D.F.V. James, A. Gilchrist, A.G. White, Experimental demonstration of a compiled version of shor's algorithm with quantum entanglement. Phys. Rev. Lett. **99**(25), 250505 (2007)
6. M.A. Nielsen, I.L. Chuang, *Quantum Computation and Quantum Information* (Cambridge University Press, Cambridge, 2000)
7. D. Beckman, A.N. Chari, S. Devabhaktuni, J. Preskill, Efficient networks for quantum factoring. Phys. Rev. A **54**(2), 1034–1063 (1996)
8. T.C. Ralph, N.K. Langford, T.B. Bell, A.G. White, Linear optical controlled-not gate in the coincidence basis. Phys. Rev. A **65**, 062324 (2002)
9. H.F. Hofmann, S. Takeuchi, Quantum phase gate for photonic qubits using only beam splitters and postselection. Phys. Rev. A **66**, 024308 (2001)
10. J.L. O'Brien, G.J. Pryde, A.G. White, T.C. Ralph, D. Branning, Demonstration of an all-optical quantum controlled-NOT gate. Nature **426**(6964), 264–267 (2003)
11. J.L. O'Brien, G.J. Pryde, A. Gilchrist, D.F.V. James, N.K. Langford, T.C. Ralph, A.G. White, Quantum process tomography of a controlled- NOT gate. Phys. Rev. Lett. **93**(8), 080502 (2004)
12. N.K. Langford, T.J. Weinhold, R. Prevedel, K.J. Resch, A. Gilchrist, J.L. O'Brien, G.J. Pryde, A.G. White, Demonstration of a simple entangling optical gate and its use in bell-state analysis. Phys. Rev. Lett. **95**(21), 210504 (2005)
13. N. Kiesel, C. Schmid, U. Weber, R. Ursin, H. Weinfurter, Linear optics controlled-phase gate made simple. Phys. Rev. Lett. **95**(21), 210505 (2005)
14. R. Okamoto, H.F. Hofmann, S. Takeuchi, K. Sasaki, Demonstration of an optical quantum controlled-not gate without path interference. Phys. Rev. Lett. **95**(21), 210506 (2005)
15. A.S. Clark, J. Fulconis, J.G. Rarity, W.J. Wadsworth, J.L. O'Brien, All-optical-fiber polarization-based quantum logic gate, Phys. Rev. A **79**, 030303(R) (2009)
16. A. Politi, M.J. Cryan, J.G. Rarity, S. Yu, J.L. O'Brien, Silica-on-silicon waveguide quantum circuits. Science **320**(5876), 646–649 (2008)

17. A. Laing, A. Peruzzo, A. Politi, M. Rodas Verde, M. Halder, T.C. Ralph, M.G. Thompson, J.L. O'Brien, High-fidelity operation of quantum photonic circuits. Appl. Phys. Lett. **97**, 211109 (2010)
18. A. Politi, J.C.F. Matthews, M.G. Thompson, J.L. O'Brien, Integrated quantum photonics. IEEE J. Sel. Top. Quantum Electron. **15**, 1673–1684 (2009)
19. J.L. O'Brien, Optical quantum computing. Science **318**(5856), 1567–1570 (2007)
20. C.K. Hong, Z.Y. Ou, L. Mandel, Measurement of subpicosecond time intervals between two photons by interference. Phys. Rev. Lett. **59**(18), 2044–2046 (1987)

Chapter 5
Quantum Interference in a Waveguide Interferometer

5.1 Introduction

Quantum-enhanced technologies require methods to precisely prepare and control quantum systems including performing state and process tomography—verifying coherent quantum operation [2]—and the optical switching required for feed-forward for optical quantum computing [3]. From a practical point of view, experimental quantum physics can greatly benefit from reconfigurability, allowing realisation of multiple quantum optical circuits using a single nested circuit of interferometers and appropriate phase shifters capable of implementing arbitrary unitary operation acting on an arbitrary number N of optical modes [4]. The key to reconfigurable quantum optical circuits is to control the parameters that describe the space of unitary operations acting on N modes, SU(N) [5]. The simplest reconfigurable quantum circuit is equivalent to a balanced interferometer that realises a subset of SU(2) and is the basic circuit component from which to construct arbitrary reconfigurable circuitry [4].

In this chapter, we highlight the potential of integrated optics for realising reconfigurable circuits. We use a Mach Zehnder interferometer—integrated on a waveguide chip with optical phase control based on a resistive heating element—as a reconfigurable reflectivity beamsplitter to continuously and accurately tune Hong–Ou–Mandel interference visibility, yielding a maximum visibility of $98.2 \pm 0.9\,\%$. We use the circuit to manipulate a single qubit—encoded on a single photon in two paths—observing an interference contrast of $98.2 \pm 0.3\,\%$. Finally, we generate and manipulate post-selected photon-number path entanglement—relevant to quantum metrology—within the chip, observing 2- and 4-photon super-resolved interference fringes with respective contrasts $97.2 \pm 0.4\,\%$ and $92 \pm 4\,\%$ sufficient to beat the standard quantum limit.

Some of the results reported in this chapter and an abbreviated discussion thereof were published as Ref. [1].

J. C. F. Matthews, *Multi-Photon Quantum Information Science and Technology in Integrated Optics*, Springer Theses, DOI: 10.1007/978-3-642-32870-1_5, © Springer-Verlag Berlin Heidelberg 2013

5.2 Integrated Mach Zehnder Interferometer with a Resistive Heating Element

Interferometers consist of a means of splitting and recombining optical modes (typically beamsplitters) and an accurate and stable means for controlling optical path lengths on a sub-wavelength scale, equivalent to a phase shift or $Z(\theta)$ operation. The whole circuit is then often required to be kept stable on a sub-wavelength scale. As with path encoding in general, accurate realisation of $Z(\theta)$ operations requires passively stable architectures such as beam displacers [6] and displaced Sagnac interferometers [7] or active stabilisation techniques (used for example in [8]).

The inherent stability offered by monolithic integrated waveguide optics provides straightforward realisation of Mach Zehnder interferometers with phase shifts that can be realised using for example resistive heating elements as reported here. Such an implementation has the benefits of a miniaturised architecture (the interferometer discussed here is on the mm scale), independent operation to the environment and a route to realising fully reconfigurable and nested interferometers acting on many modes, as required in proposal [4]. The schematic in Fig. 5.1a is of a Silica-on-Silicon (see Chap. 2) Mach Zehnder interferometer constructed from two lithographically patterned directional couplers and an electrically controlled resistive heating element that locally heats part of the waveguide circuit. The resistive heating element was fabricated as a final metal layer lithographically patterned on the top of the devices to form resistive elements (R) and the metal connections and contact pads (p_1 and p_2) shown in Fig. 5.1a, b.

Applying a voltage between p_1 and p_2 allows the current in R to generate heat which dissipates into the local structure and raises the temperature T of the cladding and core of the waveguide section directly beneath R. The change in refractive index n of silica is given to first approximation [9] by $dn/dT = 10^{-5}/\mathrm{K}$ (independent of compositional variation in the structure) which in turn alters the mode group index of the light confined in the waveguide beneath R. The heat generated inside R dissipates through the depth of the structure to the silicon substrate which acts as a heat sink. The element was designed to provide a continuously variable phase shift that includes the range $\phi \in [-\pi/2, \pi/2]$ for wavelengths in the range 780–810 nm and for operation at room temperature.

5.3 Mach Zehnder Interferometer Matrix Representation

The reported interferometer consists of two $\eta = 1/2$ reflectivity beamsplitters, which is modelled as in previous chapters with complex Hadamard matrix

$$H_c \doteq \frac{1}{\sqrt{2}} \begin{pmatrix} 1 & i \\ i & 1 \end{pmatrix} \tag{5.1}$$

equal to the square-root of a NOT or σ_x operation.

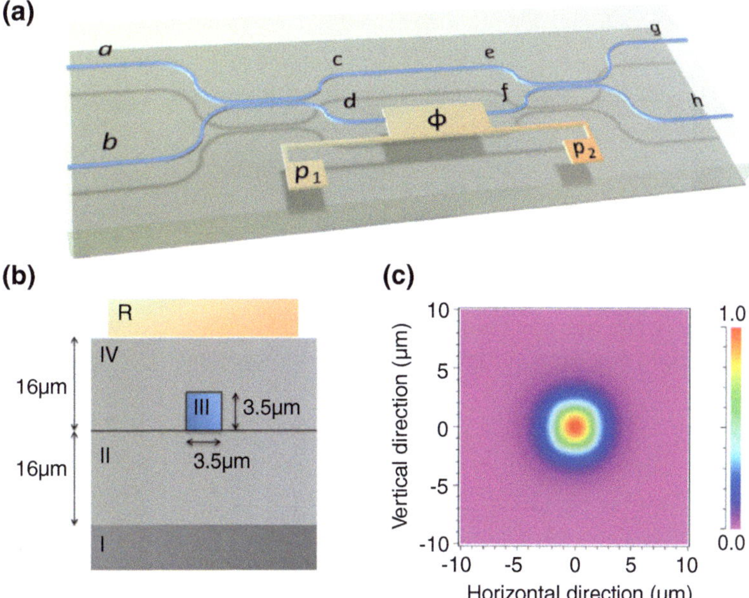

Fig. 5.1 Controlling quantum states of light in waveguide circuits. **a** Schematic of an integrated Mach Zehnder interferometer realised with waveguide and resistive heating elements (not to scale). The beamsplitters are realised with 50:50 ratio directional couplers while the controllable phase shifter is realised by a resistive heater placed above one arm of the interferometer. **b** Cross section illustrating the structure of one waveguide located beneath a resistive heater. **c** The simulation of the intensity profile single mode guided in the silica waveguide for the operated wavelength of 780 nm. Figure reproduced from Ref. [1]

Relative optical phase shifts between two optical modes are modelled by the operation $e^{i\frac{\phi}{2}\sigma_z} = Z(\phi)$; for a single photon path encoded qubit this is equal to a ϕ- rotation around the Z-axis of the Bloch sphere and is represented by the matrix

$$Z(\phi) \doteq \begin{pmatrix} e^{i\phi/2} & 0 \\ 0 & e^{-i\phi/2} \end{pmatrix} \tag{5.2}$$

It follows that H_c is related by two phase shifts of $e^{i\frac{3\pi}{4}\sigma_z}$ to the real valued Hadamard operator H according to

$$H_c = i e^{i\frac{3\pi}{4}\sigma_z} H e^{i\frac{3\pi}{4}\sigma_z} \tag{5.3}$$

Computing the matrix product $H_c Z(\phi) H_c$, we see that a directional coupler-based Mach Zehnder interferometer (Fig. 5.1a) is represented by the matrix

$$U_{mz}(\phi) = H_c e^{i\frac{\phi}{2}\sigma_z} H_c \doteq -i \begin{pmatrix} \sin\frac{\phi}{2} & \cos\frac{\phi}{2} \\ \cos\frac{\phi}{2} & -\sin\frac{\phi}{2} \end{pmatrix} \tag{5.4}$$

which is equivalent to the operation $X(\theta) = e^{i\frac{\theta}{2}\sigma_x}$ according to

$$U_{mz}(\phi) = \left(ie^{i\frac{3\pi}{4}\sigma_z}He^{i\frac{3\pi}{4}\sigma_z}\right)e^{i\frac{\phi}{2}\sigma_z}\left(ie^{i\frac{3\pi}{4}\sigma_z}He^{i\frac{3\pi}{4}\sigma_z}\right) \tag{5.5}$$

$$= -e^{i\frac{3\pi}{4}\sigma_z}He^{i(\frac{\phi}{2}+\frac{3\pi}{2})\sigma_z}He^{i\frac{3\pi}{4}\sigma_z} \tag{5.6}$$

$$= -e^{i\frac{3\pi}{4}\sigma_z}e^{\{i(\frac{\phi-\pi}{2})\sigma_x\}}e^{i\frac{3\pi}{4}\sigma_z} \tag{5.7}$$

where we have used the relations $X(\theta) = HZ(\theta)H$ and $Z(\theta_1)Z(\theta_2) = Z(\theta_2)Z(\theta_1) = Z(\theta_1 + \theta_2)$.

Optically, we can interpret the operation of the circuit in Fig. 5.1a to act on a single photon input into mode a according to the three-step process

$$a_a^\dagger |0\rangle \xrightarrow{H_c} \frac{1}{\sqrt{2}}\left(a_c^\dagger + ia_d^\dagger\right)|0\rangle \tag{5.8}$$

$$\xrightarrow{Z(\phi)} \frac{1}{\sqrt{2}}\left(a_e^\dagger + ie^{i\phi}a_f^\dagger\right)|0\rangle \tag{5.9}$$

$$\xrightarrow{H_c} -i\left(\sin\frac{\phi}{2}a_g^\dagger + \cos\frac{\phi}{2}e^{i\phi}a_h^\dagger\right)|0\rangle \tag{5.10}$$

which (5.8) creates a superposition of the single input across modes c and d; (5.9) varies the phase of the superposition by ϕ at e and f; (5.10) before recombination of the state at the second directional coupler. Similarly, a single photon input into mode b is transformed according to

$$a_b^\dagger |0\rangle \xrightarrow{U_{mz}} -i\left(\cos\frac{\phi}{2}a_g^\dagger - \sin\frac{\phi}{2}e^{i\phi}a_h^\dagger\right)|0\rangle \tag{5.11}$$

Addition of a second phase shift at the output of the interferometer would allow arbitrary single qubit state generation, encoded across the two modes; application of the inverse circuit (using the circuit in the reverse direction, for example) would provide a means for arbitrary projective measurement. Addition of a third phase shift would provide a means for realising any two mode unitary, since combining a circuit equivalent to an arbitrary $X(\theta)$ rotation together with two further Z rotations in the product

$$Z(\gamma)X(\beta)Z(\alpha) = e^{i\gamma\sigma_z}e^{i\beta\sigma_x}e^{i\alpha\sigma_z} \tag{5.12}$$

realises an arbitrary unitary rotation acting on two modes (an arbitrary member of SU(2)), via the Euler angle construction [5, 10] parameterized by the real angles $0 \le \alpha, \gamma \le \pi$ and $0 \le \beta \le \pi/2$. In general, a Mach Zehnder interferometer is the basis for reconfigurable circuits that can realise any member of SU(N) [4], as illustrated in Fig. 5.2 for $N = 7$.

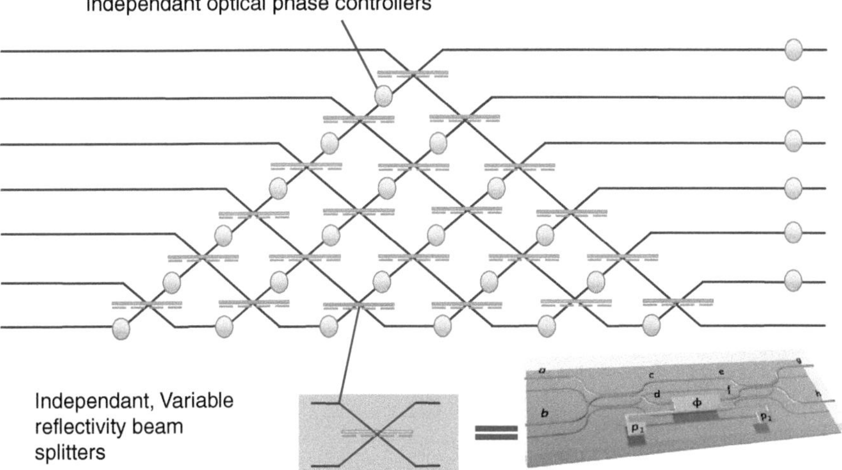

Fig. 5.2 The Reck et al. scheme [4] for realising an arbitrary unitary acting on $N = 7$ modes, illustrating how a variable beam splitter based Mach Zehnder interferometer can be used as a core component

5.4 Hong–Ou–Mandel Interference in a Reconfigurable Waveguide Beam Splitter

In general, photonic circuits such as optical entangling logic gates—[3, 6, 11] for example and as discussed in Chap. 4 [12]—are comprised of a number of different reflectivity couplers which can be adequately realised with fixed reflectivity coupler. Semi-reconfigurable photonic circuits, including routing of photons [3], and fully-reconfigurable circuits [4] can be realized by combining such variable η devices. In Chap. 3 we observed Hong–Ou–Mandel quantum interference in a range of separate directional couplers with varying reflectivity η, illustrating the effect coupling ratio has on quantum interference. With a Mach Zehnder interferometer it is possible to perform the same series of experiments with a single reconfigurable device [13].

The matrix representation of a Mach Zehnder interferometer, given in Eq. (5.4) is exactly that of a single beam splitter with variable reflectivity

$$\eta = \sin^2 \frac{\phi}{2} \tag{5.13}$$

On varying the parameter ϕ, we performed multiple Hong–Ou–Mandel experiments [14] using the device shown in Fig. 5.1 for a range of beamsplitter reflectivities η. Degenerate photon pairs from a low power pulsed SPDC source (see Chap. 2) were launched into inputs a and b of the device, with an off-chip optical delay used to scan through the photons relative arrival time. The resulting simultaneous detection rate $P_{g,h}$ of a single photon at both outputs g and h was then recorded. Each experiment

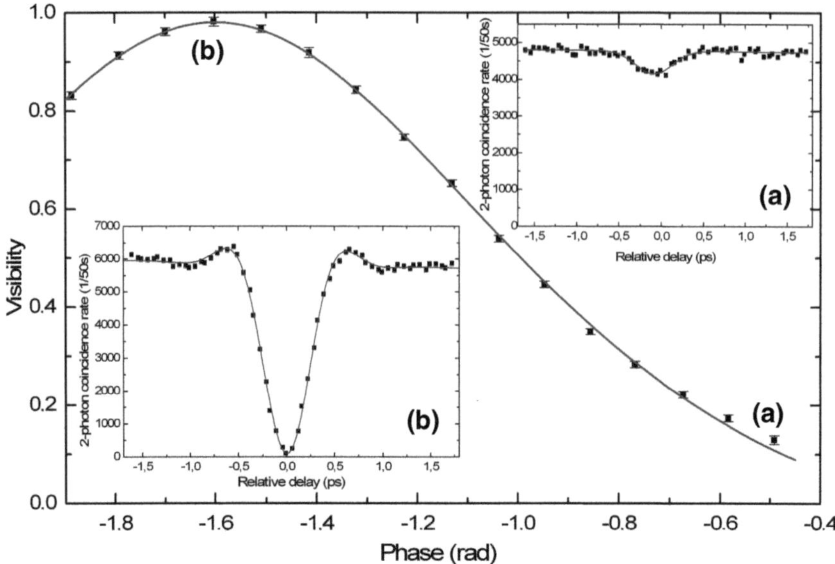

Fig. 5.3 2-photon quantum interference in a reconfigurable device. *Main panel* visibility of the Hong–Ou–Mandel experiment performed using the integrated MZ interferometer as a continuously variable beam splitter with effective reflectivity $\eta = \cos^2(\phi/2)$. The *solid line* is a theoretical fit that corresponds to Eq. (5.14), with a small phase-offset and a small amount of mode-mismatch as the only two free parameters. *Error bars* are from confidence intervals on the best-fit. *Inset* (**a**) High visibility 2-photon interference. *Inset* (**b**) Low visibility 2-photon interference. Both *inset plots* are displayed, as discussed in Chap. 3, as a plot of 2-photon rate versus the relative optical delay between the interfering photons and fitted with a function that takes into account the non Gaussian shape of the interference filter used in the experiment; error bars for each *inset* are given by Poissonian statistics. Figure reproduced from Ref. [1]

resulted in a characteristic quantum interference "dip" in $P_{g,h}$, centred around zero delay, for which we recorded the visibility (given by $V = (P_{max} - P_{min})/P_{max}$) for a range of η. As we saw in Chap. 3, ideal quantum interference is dependent upon beamsplitter reflectivity according to

$$V_{ideal} = \frac{2\eta(1-\eta)}{1-2\eta+2\eta^2} = \frac{2\sin^2\phi}{3+\cos2\phi} \tag{5.14}$$

Figure 5.3 plots the measured quantum interference visibility observed for different values of ϕ (and therefore η). The insets of Fig. 5.3 show two examples of the raw data used to generate this curve: (a) $\phi = -0.49 \pm 0.01$ radians, $V = 0.129 \pm 0.009$ and (b) $\phi = -1.602 \pm 0.01$ radians, $V = 0.982 \pm 0.009$. The average relative visibility $V_{rel} = V/V_{ideal}$ for all of the data in Fig. 5.3 is $\overline{V_{rel}} = 0.980 \pm 0.003$.

5.5 Manipulation of Multi-Photon Entanglement: NOON State Manipulation for On-Chip Quantum Metrology

Sub-wavelength sensitivity makes optical interferometry one of the most powerful precision measurement tools available to modern science and technology [15], with applications from microscopy to gravity wave detection [16, 17]. However, the use of classical states of light limits the phase precision $\Delta\phi$ of such measurements to the shot noise, or standard quantum limit (SQL): $\Delta\phi \cong 1/\sqrt{N}$, where N is the average number of sensing photons passing through the measurement apparatus. Quantum states of light—entangled states of photon number across the two paths of an interferometer for example—enable precision better than the SQL [18]. Much interest has therefore been generated around such states including the so-called "NOON state" [19].

Entangled states of $M + N$ photons across two optical modes x and y of the form

$$|N :: M\rangle^{\phi}_{x,y} = \tfrac{1}{\sqrt{2}}(|N\rangle_x |M\rangle_y + e^{i\phi}|M\rangle_x |N\rangle_y) \tag{5.15}$$

can be used to increase the frequency of interference fringes by a factor of $|N - M|$ without reducing fringe contrast, thereby increasing precision in estimating an unknown change in phase $\Delta\phi$ [20]. The canonical example is the NOON state ($M = 0$), which enables the ultimate precision $\Delta\phi \cong 1/N$—the Heisenberg limit [19].

Multi-photon interference, surpassing the SQL in some cases, has been observed with post-selected two-[21] three-[22] and four-photon states [7, 23], and with two-photon loss-tolerant non-maximally entangled states [24]. For many precision measurement applications, it is useful to encode the state using two or more spatial modes. Stability required for such encoding can be readily achieved in compact integrated quantum photonic devices [11, 25], as demonstrated by two-[1, 26] and four-photon [1] interference (Ref. [1] is the focus of this chapter).

The most trivial NOON state to observe is of a single photon: $|1 :: 0\rangle$. The sinusoidal interference pattern of a one-photon NOON state interference fringe is identical to that of a classical light, and can therefore offer by itself no quantum advantage for metrology. However, the resulting single photon interference pattern arising from manipulating the 1-photon NOON state $|1 :: 0\rangle$ within the reported device (Fig. 5.1) does correspond to the manipulation of a single photon qubit encoded across two optical modes,[1] when launching single photons into input a and controlling ϕ. The state $|1 :: 0\rangle^{\pi/2}_{c,d} = \tfrac{1}{\sqrt{2}}(|1\rangle_c |0\rangle_d + i |0\rangle_c |1\rangle_d)$ is generated after the first directional coupler as described in Eq. (5.8). The state is then manipulated using the heating element (Eq. (5.9)) before the two modes e and f interfere at the second directional coupler (Eq. (5.10)), leading to ideal photon detection rates P_g and P_h at outputs g

---

[1] For qubit encoding we define the Fock state $|1\rangle_i |0\rangle_j = |0\rangle$ as the logical 0 and the Fock state $|0\rangle_i |1\rangle_j = |1\rangle$ as the logical 1 for $(i, j) = (a, b), (c, d), (e, f), (g, h)$ throughout various stages of the circuit.

and h respectively: $P_g = 1 - P_h = \frac{1}{2}[1 - \cos(\phi)]$, yielding sinusoidal interference fringes with a period of 2π. The observed fringes are given in Fig. 5.4a and have contrast $C = 0.982 \pm 0.003$, defined for a given detection rate P by

$$C = \frac{P_{max} - P_{min}}{P_{max} + P_{min}} \tag{5.16}$$

Assuming no mixture or complex phase is introduced, the average fidelity F between the measured and ideal output state $(\sin(\phi/2)|1\rangle_g |0\rangle_h + i \cos(\phi/2)|0\rangle_g |1\rangle_h)$ is computed from the data in Fig. 5.4a; averaging over the range $\phi \in [-\pi/2, \pi/2]$ we observe $\overline{F} = 0.99984 \pm 0.00004$.

Ideally, the two photon NOON state $|2 :: 0\rangle$ is generated from Hong–Ou–Mandel interference of two degenerate photon pairs at the first beamsplitter of a Mach Zehnder interferometer [13, 21], as shown by Eq. (3.5). Following Eqs. (5.8)–(5.10), a degenerate photon pair state evolves in our device according to the following three stages (denoted by a '$\longrightarrow$')

$$a_a^\dagger a_b^\dagger |0\rangle \xrightarrow{H_c} \frac{i}{2}\left(a_c^{\dagger 2} + a_d^{\dagger 2}\right)|0\rangle \tag{5.17}$$

$$\xrightarrow{Z(\phi)} \frac{i e^{i\phi}}{2}\left(a_e^{\dagger 2} + e^{-i2\phi} a_f^{\dagger 2}\right)|0\rangle \tag{5.18}$$

$$\xrightarrow{H_c} \left[-\cos\phi\, a_g^\dagger a_h^\dagger - \sin\phi \left(\frac{a_g^{\dagger 2} - a_h^{\dagger 2}}{2} \right) \right]|0\rangle \tag{5.19}$$

$$= -\cos\phi\, |1\rangle_g |1\rangle_h - \sin\phi\, |2 :: 0\rangle_{g,h}^\pi \tag{5.20}$$

Therefore, by varying the phase of the device by ϕ, the relative phase of the NOON state inside varies according to 2ϕ as shown in Eq. (5.18). This "$\lambda/2$" interference effect manifests itself in the probability of detecting the output state $|1\rangle_g |1\rangle_h$, for example. Simultaneous detection of a single photon at each output g and h were observed as a function of ϕ, yielding an interference fringe described by $P_{g,h} = \cos^2\phi = \frac{1}{2}(1 + \cos 2\phi)$ with period π—half the period of the 1-photon interference fringes. The 2-photon interference fringe shown in Fig. 5.4b plots the simultaneous detection rate as a function of the phase ϕ. The contrast is $C = 0.972 \pm 0.004$, which is greater than the threshold $C_{th} = 1/\sqrt{2}$ required to beat the standard quantum limit for 2-photon interference [7, 20, 22, 27].

Four photon NOON states are not so straight-forward to generate. However, the dynamics of the four photon NOON state can be observed in post-selection. On inputting the four photon state $|2\rangle_a |2\rangle_b$, the state evolves through the interferometer according to the following stages (each step is presented in both creation operator and Fock state notation for clarity)

$$\frac{a_a^{\dagger 2} a_b^{\dagger 2}}{2}|0\rangle \xrightarrow{H_c} \left(-\frac{1}{4} a_c^{\dagger 2} a_d^{\dagger 2} - \frac{1}{8}\left(a_c^{\dagger 4} + a_d^{\dagger 4}\right) \right)|0\rangle \tag{5.21}$$

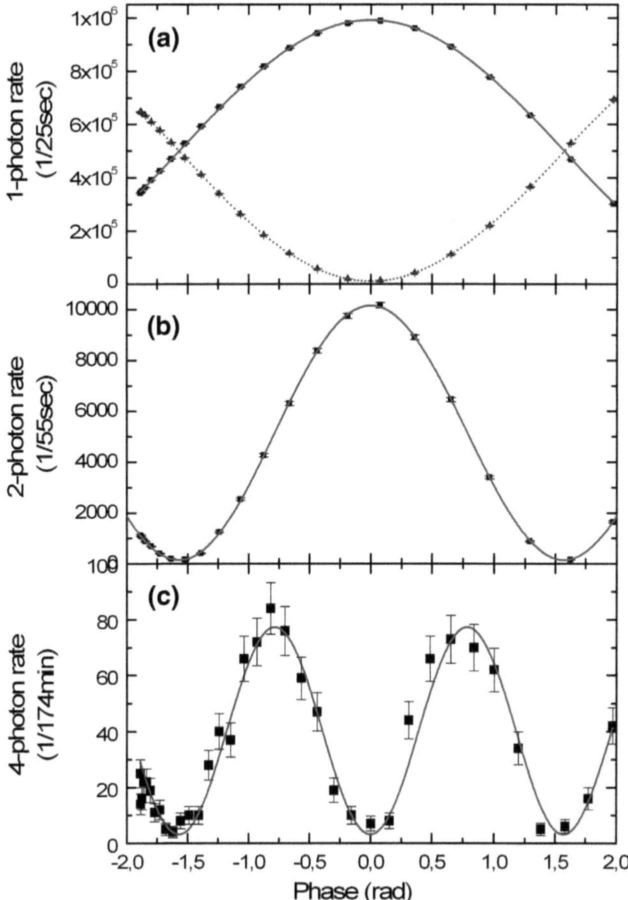

Fig. 5.4 Integrated quantum metrology. **a** 1-photon count rates at the outputs g (*blue triangle data points, dotted fit*) and h (*black square data points, solid fit*) as the phase ϕ is varied on inputting the 1-photon state $|1\rangle_a |0\rangle_b$. **b** 2-photon coincidental detection rate between the outputs g and h when inputting the 2-photon state $|1\rangle_a |1\rangle_b$ and varying the phase ϕ. **c** 4-photon coincidental detection rate of the output state $|3\rangle_g |1\rangle_h$ when inputting the 4-photon state $|2\rangle_a |2\rangle_b$. Error bars are given by Poissonian statistics. Figure reproduced from Ref. [1] (Color figure online)

$$= -\frac{1}{\sqrt{4}} |2\rangle_c |2\rangle_d - \sqrt{\frac{3}{4}} \left(\frac{|4\rangle_c |0\rangle_d + |0\rangle_c |4\rangle_d}{\sqrt{2}} \right) \tag{5.22}$$

$$\xrightarrow{Z(\phi)} \left(-\frac{1}{4} a_e^{\dagger 2} a_f^{\dagger 2} - \frac{e^{i2\phi}}{8} a_e^{\dagger 4} - \frac{e^{-i2\phi}}{8} a_f^{\dagger 4} \right) |0\rangle \tag{5.23}$$

$$= -\frac{1}{\sqrt{4}} |2\rangle_e |2\rangle_f - e^{i2\phi} \sqrt{\frac{3}{4}} \left(\frac{|4\rangle_e |0\rangle_f + e^{-i4\phi} |0\rangle_e |4\rangle_f}{\sqrt{2}} \right) \tag{5.24}$$

$$\xrightarrow{H_c} \left(\frac{(1 - \cos 2\phi)}{16} \left(a_g^{\dagger 4} + a_h^{\dagger 4} \right) + \frac{(1 + 3\cos 2\phi)}{8} a_g^{\dagger 2} a_h^{\dagger 2} \right.$$

$$\left. + \frac{\sin 2\phi}{4} \left(a_g^{\dagger 3} a_h^{\dagger} - a_g^{\dagger} a_h^{\dagger 3} \right) \right) |0\rangle \tag{5.25}$$

$$= \frac{\sqrt{3}}{4} (1 - \cos 2\phi) \left(\frac{|4\rangle_g |0\rangle_h + |0\rangle_g |4\rangle_h}{\sqrt{2}} \right) + \frac{(1 + 3\cos 2\phi)}{4} |2\rangle_g |2\rangle_h$$

$$+ \frac{\sqrt{3}}{2} \sin 2\phi \left(\frac{|3\rangle_g |1\rangle_h - |1\rangle_g |3\rangle_h}{\sqrt{2}} \right) \tag{5.26}$$

After the first directional coupler the four-photon Holland and Burnett state [28] is produced (Eqs. (5.21) and (5.22)) which for four photons is a superposition of the desired NOON state $|4 :: 0\rangle_{c,d}^0$ and the product state $|2\rangle_c |2\rangle_d$. The product state contains no relative phase information between modes e and f when ϕ is varied, and because this term is identical to the input state (with reduced amplitude) the effect of a second beamsplitter H_c give rise to another Holland and Burnett state. Only the $|4 :: 0\rangle_{e,f}^{-i4\phi}$ term in Eqs. (5.23) and (5.24) give rise to the term $\sin 2\phi |3 :: 1\rangle_{g,h}^{\pi}$ (in Eqs. (5.25) and (5.26)), therefore detection of either $|3\rangle_g |1\rangle_h$ or $|3\rangle_g |1\rangle_h$ allows post selection of the four photon NOON state. This approach was described in [29] and applied in [7, 20] with a displaced Sagnac architecture using bulk optical elements.

The probability of detecting either of the states $|3\rangle_e |1\rangle_f$ or $|1\rangle_e |3\rangle_f$ is given by $\frac{3}{8} (1 - \cos 4\phi)$, yielding a "$\lambda/4$" interference fringe with period $\pi/2$. Data taken from performing such an experiment in the waveguide Mach Zehnder interferometer is plotted as the 4-photon interference fringe shown in Fig. 5.4c, which plots the rate of simultaneous detection of four photons corresponding to the state $|3\rangle_g |1\rangle_h$ (by cascading three detectors using 1×2 fibre-beam splitters at the output g) against the phase ϕ. The contrast of this 4-photon interference is $C = 0.92 \pm 0.04$, which is greater than the threshold $C_{th} = 1/\sqrt{4}$ to in principle beat the standard quantum limit [20].

The 4-photon interference experiment was repeated using a higher pump power from the Ti:Sapphire laser pumping the down-conversion process (see Chap. 2) with the resulting data plotted in Fig. 5.5. This was performed to confirm with a higher accuracy the reduced de Broglie wavelength by obtaining lower error bars for the experiment. As the power of the pump is increased, the 4-photon production rate increases, but by the same argument, the production of 6-photons start to be non-negligible. This reduces the contrast of the measured fringe ($C = 83.1 \pm 1.5\%$), since losses and avalanche detectors that cannot resolve photon number give rise to spurious counts of the $|3\rangle_g |1\rangle_h$ state. Despite the post-selecting detection scheme and the higher power, the measured contrast is still in principle sufficient to beat the shot noise limit.

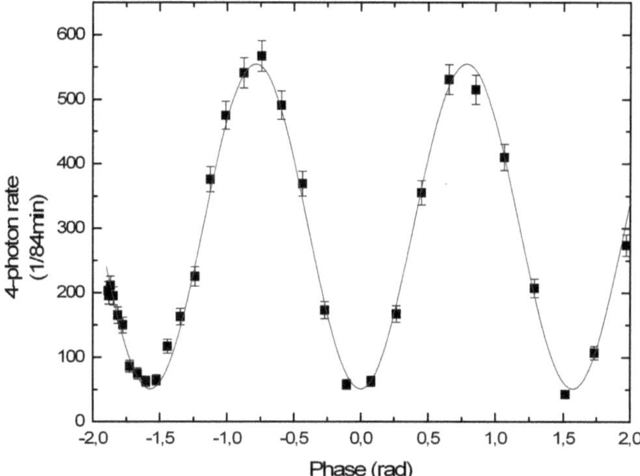

Fig. 5.5 4-photon detection rate of the output state $|3\rangle_g\,|1\rangle_h$ for higher laser power pumping the down-conversion process than for Fig. 5.4c. Figure reproduced from supplementary information of Ref. [1]

5.6 Multi-Photon Detection

The 1-, 2- and 4-photon detection schemes for the interference fringe experiments shown in Figs. 5.4 and 5.5 are shown in Fig. 5.6. The 1-photon interference fringes (Fig. 5.4a) are observed using one output of the Type-I spontaneous parametric photon source (see Chap. 2) coupled to input a of the waveguide circuit; measurement with respect to phase ϕ is conducted by coupling each of the outputs g and h to fibre coupled single photon counting modules labeled A and B and monitoring the respective single photon count rates.

The 2-photon "$\lambda/2$" interference fringes are observed by measuring the twofold coincidental photon detection rate across single photon counting modules A and B (Fig. 5.6b). The same 2-photon setup and detection scheme is used for the multiple Hong–Ou–Mandel experiments reported in Fig. 5.3.

The 4-photon interference fringes are observed using 4-photons emitted in two modes from the pulsed spontaneous parametric photon source (see Chap. 2) and coupled into modes a and b. The "$\lambda/4$" fringe is measured by detecting either the $|3\rangle_g\,|1\rangle_h$ or $|1\rangle_g\,|3\rangle_h$ Fock states; the latter, for example, is detected non-deterministically by cascading single photon counting modules B, C and D as shown with two fibre splitters and monitoring the fourfold coincidental photon detection rate across A, B, C and D. Each fringe is plotted using the phase voltage relation derived from the resistive heater calibration in the next section.

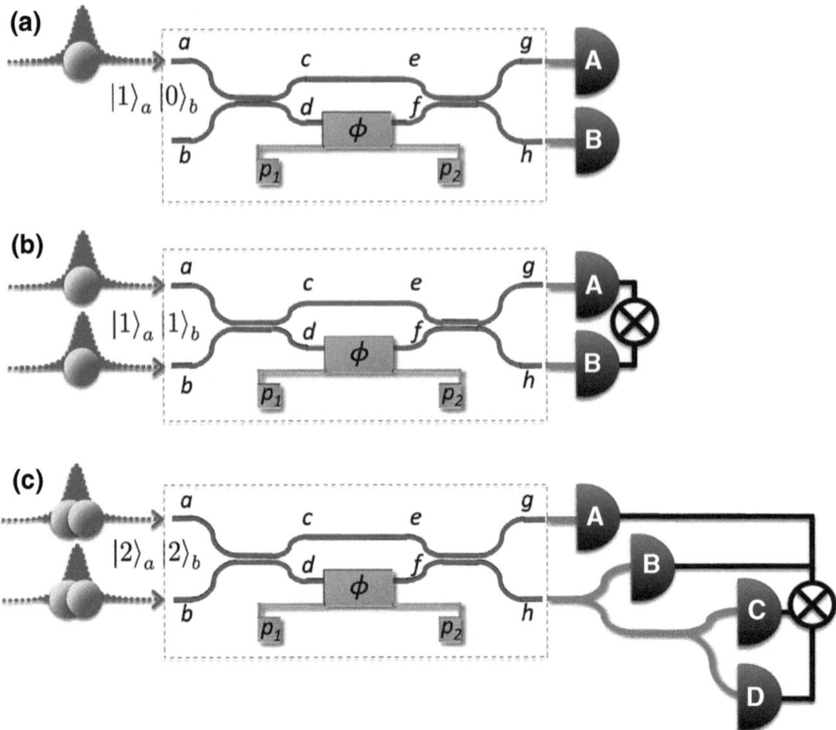

Fig. 5.6 Setup for multi-photon detection for the interference fringe experiments shown in Figs. 5.4 and 5.5. **a** The 1-photon interference fringe excitation and detection scheme. **b** The 2-photon "$\lambda/2$" interference fringe excitation and detection scheme. **c** The 4-photon "$\lambda/4$" interference fringe excitation and detection scheme. Figure from supplementary information of Ref. [1]

5.7 Phase Control Calibration

Before performing the experiments reported in this chapter, the electrically driven phase shift inside the device had to be calibrated in order to determine the non-linear phase-voltage relationship $\phi(V)$ determined at the time of chip fabrication. For this reason each such device has to be initially calibrated. Calibration of $\phi(V)$ modulo π harnesses the reduced de Broglie wavelength [7, 22, 23] of 2-photon interference [13, 21, 27, 30–33] to more widely sample the pattern of phase dependent interference of the maximally path entangled state $\frac{1}{\sqrt{2}}\left(|2\rangle_e |0\rangle_f + e^{i2\phi(V)} |0\rangle_e |2\rangle_f\right)$ generated inside the interferometer (Sect. 5.5). The phase shift ambiguity is later corrected to modulo 2π by direct comparison to well-known 1-photon "classical" interference.

Applied phase ϕ is proportional, to first approximation, to the power dissipated by the resistor. This translates into a quadratic relation between the applied voltage and the phase. To take into account deviations form the ideal case, mainly due to a

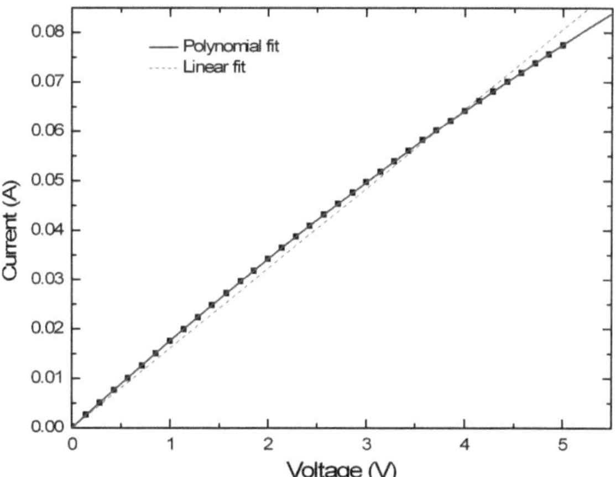

Fig. 5.7 Current-voltage relation of the resistive heater. To highlight the non-ohmic relation, we show a linear and a polynomial best fit. Figure from supplementary information of Ref. [1]

non-ohmic current-voltage relation as the temperature changes (as shown in Fig. 5.7), we fixed the form of the $\phi(V)$ relation as

$$\phi(V) = \alpha + \beta V^2 + \gamma V^3 + \delta V^4 \tag{5.27}$$

where the V^3 term is related to non-ohmic behavior of the resistor, and the V^4 term is a combination of non-ohmic relation and the expansion of $\phi(V)$ in even powers of V.

Figure 5.8 plots the two-photon coincidence rate $P_{g,h}$ at the output of the chip on inputting $|1\rangle_a |1\rangle_b$, as a function of voltage V applied across p_1 and p_2. The parameters computed from the calibration process for $\phi(V)$ were found by means of best-fit of the data presented in Fig. 5.8 and reported in Table 5.1, which we use throughout our analysis of the 1-, 2- and 4-photon interference experiments as well as for varying the reflectivity in 2-photon Hong–Ou–Mandel experiments reported in the proceeding sections of this chapter. The resulting relationship is plotted in the inset of Fig. 5.8.

Stability of the phase applied in the Mach Zehnder interferometer was tested using single photons. Figure 5.9 shows the results of this experiment and corresponds to probability of detecting a photon in mode h when sending single photons in input a as a function of time, a voltage of 1.4 V was applied across resistive element. The probability remained almost constant for more than six hours. The small deviation is attributed to the different evolution of the coupling from the waveguides g and h to the fiber array that collect the photons at the output of the circuit. However, different evolution of the coupling efficiencies (loss) does not lower the quantum mechanical performance of the device.

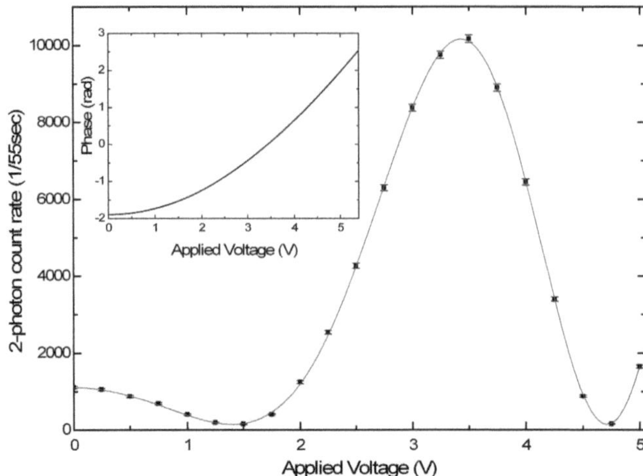

Fig. 5.8 The 2-photon interference fringe used to calibrate the voltage-controlled phase shift $\phi(V)$. *Main panel* the 2-photon interference pattern generated from simultaneous detection of a single photon at both outputs g and h as the voltage applied across the device was varied between 0 and 5 V. Error bars are given by Poissonian statistics. *Inset* plot of the phase-voltage relationship determined from this calibration. Figure from Ref. [1]

Table 5.1 Values of the parameters obtained from the best fit in the calibration of $\phi(V)$ given in Eq. (5.27)

Parameter	Value	Error
α	-1.887	0.006
β	0.157	0.005
γ	0.0045	0.002
δ	-0.0001	0.0002

5.8 Discussion

The reconfigurable quantum circuit demonstrated here could be used as the fundamental element to build a large-scale circuit capable of implementing any unitary operation on many waveguides; this circuit has already been demonstrated as a fundamental building block for a reconfigurable two-qubit processor [34]. In classical optics, examples of a thermal-based optical switching include a demonstration of a 32×32 waveguide switch consisting of 2048 individual components [35]. Implementing an arbitrary unitary (Ref. [4] and Fig. 5.2) on this number of modes would require a comparable number of components which is clearly far beyond any practical implementation with conventional bulk optics.

The millisecond timescales available with thermal switching, and the phase control reported in this chapter, imply thermal phase manipulation is suitable for reconfigurable single photon circuits, for state preparation, quantum measurement and

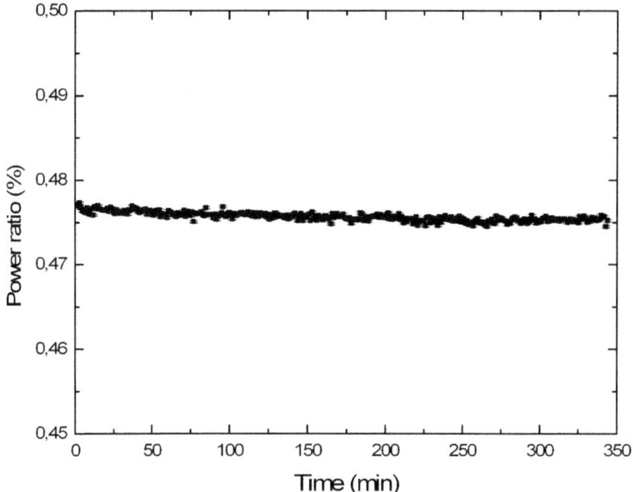

Fig. 5.9 Probability of detecting a photon in mode h when sending a single photon in input a as a function of time. To see the stability of the phase, the probability axis is zoomed in on the range (0.45, 0.5). Figure from supplementary information of Ref. [1]

for integrated quantum metrology [18]. In addition to the single demonstrations presented here, these devices may be used for other quantum states of light. In particular, phase control will be particularly important for homodyne detection required for phase estimation [36] and adaptive measurements [37] with squeezed states of light, as well as integration with unknown shifts, such as using opto-fluidics [38], for quantum-enhanced phase estimation.

Other applications demanding fast switching, such as adaptive circuits for quantum control and feedforward, will require sub-nanosecond switching, which is possible using electro-optic materials such as $LiNbO_3$, used to make modulators operating at rate of $O(10)GHz$ [39].

References

1. J.C.F. Matthews, A. Politi, A. Stefanov, J.L. O'Brien, Manipulation of multiphoton entanglement in waveguide quantum circuits. Nat. Photonics **3**, 346–350 (2009)
2. J.L. O'Brien, G.J. Pryde, A. Gilchrist, D.F.V. James, N.K. Langford, T.C. Ralph, A.G. White, Quantum process tomography of a controlled-NOT gate. Phys. Rev. Lett. **93**(8), 080502 (2004)
3. E. Knill, R. Laflamme, G.J. Milburn, A scheme for efficient quantum computation with linear optics. Nature **409**(6816), 46–52 (2001)
4. M. Reck, A. Zeilinger, H.J. Bernstein, P. Bertani, Experimental realization of an discrete unitary operator. Phys. Rev. Lett. **73**(1), 58–61 (1994)
5. T. Tilma, E.C.G. Sudarshan, Generalized euler angle parameterization for su(n), J. Phys. A: Math. Gen. **35**, 10 467–10 501 (2002)

6. J.L. O'Brien, G.J. Pryde, A.G. White, T.C. Ralph, D. Branning, Demonstration of an all-optical quantum controlled-not gate. Nature **426**(6964), 264–267 (2003)
7. T. Nagata, R. Okamoto, J.L. O'brien, K. Sasaki, S. Takeuchi, Beating the standard quantum limit with four-entangled photons. Science **316**(5825), 726–729 (2007)
8. C. Braig, P. Zarda, C. Kurtsiefer, H. Weinfurter, Experimental demonstration of complementarity with single photons. App. Phys. B-Lasers O **76**(2), 113–116 (2003)
9. I. Kenichi, Y. Kokubun, *Encyclopedic Handbook of Integrated Optics* (CRC Press, London, 2006)
10. J.J. Sakurai, *Modern Quantum Mechanics*, 2nd edn. (Addison Wesley, New York, 1994)
11. A. Politi, M.J. Cryan, J.G. Rarity, S. Yu, J.L. O'Brien, Silica-on-silicon waveguide quantum circuits. Science **320**(5876), 646–649 (2008)
12. M.G. Thompson, A. Politi, J.C.F. Matthews, J.L. O'Brien, Integrated waveguide circuits for optical quantum computing. IET Circ. Devices Syst. **5**, 94–102 (2011)
13. J.G. Rarity, P.R. Tapster, E. Jakeman, T. Larchuk, R.A. Campos, M.C. Teich, B.E.A. Saleh, Two-photon interference in a mach-zehnder interferometer. Phys. Rev. Lett. **65**(11), 1348–1351 (1990)
14. C.K. Hong, Z.Y. Ou, L. Mandel, Measurement of subpicosecond time intervals between two photons by interference. Phys. Rev. Lett. **59**(18), 2044–2046 (1987)
15. F. Mayinger, O. Feldmann, *Optical Measurements: Techniques and Applications*, 2nd edn. (Springer, New York, 2002)
16. A. Abramovici, W.E. Althouse, R.W.P. Drever, Y. Gursel, S. Kawamura, F.J. Raab, D. Shoemaker, L. Sievers, R.E. Spero, K.S. Thorne, R.E. Vogt, R. Weiss, S.E. Whitcomb, M.E. Zucker, Ligo: The laser interferometer gravitational-wave observatory. Science **256**(5055), 325–333 (1992)
17. K. Goda, O. Miyakawa, E.E. Mikhailov, S. Saraf, R. Adhikari, K. McKenzie, R. Ward, S. Vass, A.J. Weinstein, N. Mavalvala, A quantum-enhanced prototype gravitational-wave detector. Nat. Phys. **4**(6), 472–476 (2008)
18. V. Giovannetti, S. Lloyd, L. Maccone, Quantum-enhanced measurements: Beating the standard quantum limit. Science **306**, 1330–1336 (2004)
19. J.P. Dowling, Quantum optical metrology - the lowdown on high-N00N states. Contemp. Phys. **49**, 125–143 (2008)
20. R. Okamoto, H.F. Hofmann, T. Nagata, J.L. O'Brien, K. Sasaki, S. Takeuchi, Beating the standard quantum limit: phase super-sensitivity of n-photon interferometers. New J. Phys. **10**(7), 073033 (2008)
21. Z.Y. Ou, X.Y. Zou, L.J. Wang, L. Mandel, Experiment on nonclassical 4th-order interference. Phys. Rev. A **42**, 2957–2965 (1990)
22. M.W. Mitchell, J.S. Lundeen, A.M. Steinberg, Super-resolving phase measurements with a multiphoton entangled state. Nature **429**(6988), 161–164 (2004)
23. P. Walther, J.W. Pan, M. Aspelmeyer, R. Ursin, S. Gasparoni, A. Zeilinger, De broglie wavelength of a non-local four-photon state. Nature **429**(6988), 158–161 (2004)
24. M. Kacprowicz, R. Demkowicz-Dobrzanski, W. Wasilewski, K. Banaszek, I.A. Walmsley, Experimental quantum-enhanced estimation of a lossy phase shift. Nat. Photonics **4**, 357–360 (2010)
25. G.D. Marshall, A. Politi, J.C.F. Matthews, P. Dekker, M. Ams, M.J. Withford, J.L. O'Brien, Laser written waveguide photonic quantum circuits. Opt. Express **17**(15), 12 546–12 554 (2009)
26. B.J. Smith, D. Kundys, N. Thomas-Peter, P.G.R. Smith, I.A. Walmsley, Phase-controlled integrated photonic quantum circuits. Opt. Express **17**, 13 516–13 525 (2009)
27. K.J. Resch, K.L. Pregnell, R. Prevedel, A. Gilchrist, G.J. Pryde, J.L. O'Brien, A.G. White, Time-reversal and super-resolving phase measurements. Phys. Rev. Lett. **98**(22), 223601 (2007)
28. M.J. Holland, K. Burnett, Interferometric detection of optical phase shifts at the heisenberg limit. Phys. Rev. Lett. **71**(9), 1355–1358 (1993)
29. O. Steuernagel, de broglie wavelength reduction for a multiphoton wave packet. Phys. Rev. A **65**(3), 033820 (2002)

30. A. Kuzmich, L. Mandel, Sub-shot-noise interferometric measurements with two-photon states. Quant. Semiclass. Opt. **10**, 493 (1998)
31. E.J.S. Fonseca, C.H. Monken, S. Pádua, Measurement of the de broglie wavelength of a multiphoton wave packet. Phys. Rev. Lett. **82**(14), 2868–2871 (1999)
32. K. Edamatsu, R. Shimizu, T. Itoh, Measurement of the photonic de broglie wavelength of entangled photon pairs generated by spontaneous parametric down-conversion. Phys. Rev. Lett. **89**(21), 213601 (2002)
33. H.S. Eisenberg, J.F. Hodelin, G. Khoury, D. Bouwmeester, Multiphoton path entanglement by nonlocal bunching. Phys. Rev. Lett. **94**(9), 090502 (2005)
34. P.J. Shadbolt, M.R. Verde, A. Peruzzo, A. Politi, A. Laing, M. Lobino, J.C.F. Matthews, M.G. Thompson, J.L. O'Brien, Generating, manipulating and measurng entanglement and mixture with a reconfigurable photonic circuit. Nat. Photonics **6**, 45–59 (2012)
35. S. Sohma, T. Watanabe, N. Ooba, M. Itoh, T. Shibata, H. Takahashi, Silica-based PLC Type 32 × 32 Optical Matrix Switch. Optical Communication, 2006. ECOC 2006. European Conference on, pp. 1–2, 24–28 Sept. 2006. doi:10.1109/ECOC.2006.4801113. http://ieeexplore. ieee.org/stamp/stamp.jsp?tp=&arnumber=4801113&isnumber=4800856
36. L. Pezzé, A. Smerzi, G. Khoury, J.F. Hodelin, D. Bouwmeester, Phase detection at the quantum limit with multiphoton mach-zehnder interferometry. Phys. Rev. Lett. **99**(22), 223602 (2007)
37. D.W. Berry, H.M. Wiseman, Adaptive phase measurements for narrowband squeezed beams. Phys. Rev. A **73**(6), 063824 (2006)
38. A. Crespi, M. Lobino, J.C.F. Matthews, A. Politi, C.R. Neal, R. Ramponi, R. Osellame, J.L. O'Brien, Measuring protein concentration with entangled photons. Appl. Phys. Lett. **100**, 233704 (2012)
39. E. Wooten, K. Kissa, A. Yi-Yan, E. Murphy, D. Lafaw, P. Hallemeier, D. Maack, D. Attanasio, D. Fritz, G. McBrien, D. Bossi, A review of lithium niobate modulators for fiber-optic communications systems. IEEE J. Sel. Top. Quantum Electron. **6**(1), 69–82 (2000)

Chapter 6
Heralded NOON State Generation in Waveguide

6.1 Introduction

Chapter 5 explored the manipulation of multi-photon entangled states observed via post selection—a technique that is a powerful approach to observing quantum mechanical effects in experimental quantum information science and a method we shall use again in Chap. 8. For many applications post-selection is not sufficient to generate useful entanglement that will provide a quantum mechanical advantage over other means. A prime example is quantum metrology, where the same level of precision in measurement can be achieved as classical metrology, but with the quantum advantage of using fewer resources. This advantage is lost when many of the quantum particles used have to be excluded from the measurement via post-selection. The photons excluded in Chap. 5 via post selection still pass through the measurement apparatus, which could be detrimental in circumstances when the sample being measured is sensitive in some way to the measurement process.

In this chapter, we experimentally investigate an alternative approach proposed [2] to herald the 2- and 4-photon NOON states $|2 :: 0\rangle$ and $|4 :: 0\rangle$ in respectively four and six photon experiments, using projective measurements on ancilla photons. Incorporating the demonstration here with fast switching and optical delay may provide an approach for quantum metrology where the sample being measured is exposed only to the desired probe light via a heralded gating system. The experiments we explore use the integrated waveguide architecture in anticipation of integrating with more complex circuitry, photon sources and detectors to realise a practical quantum technology.

Some of the results reported in this chapter and an abbreviated discussion thereof were published as Ref. [1].

J. C. F. Matthews, *Multi-Photon Quantum Information Science and Technology in Integrated Optics*, Springer Theses, DOI: 10.1007/978-3-642-32870-1_6,
© Springer-Verlag Berlin Heidelberg 2013

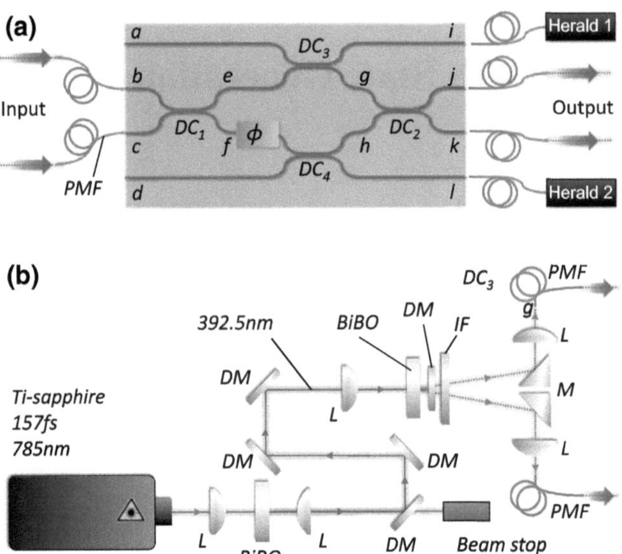

Fig. 6.1 Heralding multi-photon path-entangled states in a photonic chip. The waveguide circuit with coupling reflectivities $DC_{1,2} = 1/2$, $DC_{3,4} = 1/3$. Figure reproduced from Ref. [1]

6.2 Waveguide Circuit for Heralding Path-Number Entangled NOON States Using Projective Measurement

Entangling interactions between photons can be achieved using only linear optical circuits, additional photons and photon detection [3, 4] where a particular detection event heralds the success of a given process. In this way it is possible to generate multi-photon entangled states and indeed to efficiently perform universal, fault tolerant quantum computing [3]. There have been several examples of heralding multi-photon states for various applications (e.g. Refs. [5–12]). Generating a particular class of path entangled states for quantum metrology, including NOON states, is a particularly important example where an N-photon entangled state is heralded from $> N$ input photons and several schemes for doing this have been proposed [2, 13, 14].

The silica-on-silicon waveguide circuit shown in Fig. 6.1 is capable of heralding the two- and four-photon NOON states $|2 :: 0\rangle_{j,k}^0$ and $|4 :: 0\rangle_{j,k}^\pi$, as well as the four-photon state $|3 :: 1\rangle_{j,k}^0$, dependent upon the input state and the setting of the internal phase ϕ. The circuit consists of directional couplers DC_{1-4}, equivalent to beam splitters, used to couple photons between optical modes and for quantum interference [15] and modelled with operator

$$DC_i \doteq \begin{pmatrix} \sqrt{\eta_i} & i\sqrt{1-\eta_i} \\ i\sqrt{1-\eta_i} & \sqrt{\eta_i} \end{pmatrix} \qquad (6.1)$$

The reflectivities in the circuit are designed to have the values $\eta_1 = \eta_2 = 1/2$ and $\eta_3 = \eta_4 = 1/3$. The resistive heating element controls the relative optical phase ϕ inside the device.

6.2.1 Heralding Two Photon NOON States

The state $|2 :: 0\rangle^0_{j,k}$ can be heralded by inputting four identical photons, via polarization maintaining fibre (PMF), in the (unentangled) path encoded state $|2\rangle_b |2\rangle_c$ of four frequency degenerate photons. Quantum interference at the directional coupler DC_1 transforms the input state according to

$$|2\rangle_b |2\rangle_c \xrightarrow{DC_1} \sqrt{\frac{3}{4}} |4 :: 0\rangle^0_{e,f} + \frac{1}{\sqrt{4}} |2\rangle_e |2\rangle_f \tag{6.2}$$

After DC_3 and DC_4 this state evolves to a superposition across the four modes i, g, h and l. However, only the component $|2\rangle_e |2\rangle_f$ gives rise to terms that include $|1\rangle_i |1\rangle_l$. Subsequent detection of only one photon each in waveguides i and l, therefore project the state in modes g and h to

$$\sqrt{\frac{4}{81}} |1\rangle_i |1\rangle_g |1\rangle_h |1\rangle_l \tag{6.3}$$

The value of the variable phase ϕ is uninfluential in this case, since it can be treated as a global phase. Quantum interference [16] at the final directional coupler DC_2 yields the two photon state

$$\sqrt{\frac{4}{81}} |1\rangle_i |2 :: 0\rangle^0_{j,k} |1\rangle_l \tag{6.4}$$

corresponding to the two-photon NOON state $|2 :: 0\rangle_{j,k}$ at the output of the photonic chip. In the original reference [2], DC_3 and DC_4 both have reflectivity $\eta = 0.5$, leading to the intrinsic heralding success rate of $1/16$—i.e. the probability of detecting $|1\rangle_i |1\rangle_l$ and thereby heralding $|2 :: 0\rangle^0_{j,k}$. Note that with sufficient gating, heralding rates do not degrade the level of accuracy below the Heisenberg limit.

For a low loss regime and in the absence of higher photon number terms, the heralding of the $|1\rangle_i |1\rangle_l$ component eliminates the lower order input state $|1\rangle_b |1\rangle_c$. We note that the requirements of heralding states for quantum metrology are more relaxed than for quantum computation or cryptography. A false heralded event of the vacuum state (due for example to loss) would be detrimental for any computation. In contrast, when low photon flux is the main requirement (exposure of a measured sample to radiation is to be kept to a minimum), a false heralding event of a vacuum state will not expose the sample to radiation.

6.2.2 Heralding Four Photon Entangled States

The four-photon states $|3 :: 1\rangle^{\phi}_{j,k}$ and $|4 :: 0\rangle^{\pi}_{j,k}$ are heralded in a similar manner. On inputting six indistinguishable photons in the state $|3\rangle_b |3\rangle_c$ into the chip, quantum interference at DC_1 coherently transforms the state according to

$$|3\rangle_b |3\rangle_c \xrightarrow{DC_1} \sqrt{\frac{5}{8}} |6 :: 0\rangle^0_{e,f} + \sqrt{\frac{3}{8}} |4 :: 2\rangle^0_{e,f} \tag{6.5}$$

The couplers DC_3 and DC_4, combined with the subsequent detection of only one photon each in waveguides i and l, project the state to

$$|1\rangle_i \sqrt{\frac{4}{243}} \left(\frac{|3\rangle_g |1\rangle_h + e^{2i\phi} |1\rangle_g |3\rangle_h}{\sqrt{2}} \right) |1\rangle_l \tag{6.6}$$

The variable internal phase ϕ (controllable via a thermo-optical electrode) can be used to control this state, with the effect of non-classical interference at directional coupler DC_2 depending on the phase ϕ according to

$$|1\rangle_i \sqrt{\frac{4}{243}} \left(\sin \phi \, |4 :: 0\rangle^{\pi}_{j,k} - \cos \phi \, |3 :: 1\rangle^0_{j,k} \right) |1\rangle_l \tag{6.7}$$

The state in Eq. (6.6) can now be rotated by the electrode with a "$\lambda/2$" de Broglie wavelength on detection of the state $|4\rangle_j |0\rangle_k$. This provides an important means of testing quantum coherence within the optical circuit. With the phase set to $\phi = 0$, the state returns to $|3 :: 1\rangle^0_{j,k}$ after DC_2. With the phase set to $\phi = \pi/2$ quantum interference at DC_2 yields the four photon NOON state $|4 :: 0\rangle^{\pi/2}_{j,k}$. For $\eta = 0.5$ for both DC_3 and DC_4, the success rate of heralding $|4 :: 0\rangle^{\pi}_{j,k}$ at the output is 3/64 [2].

6.3 Heralding a Two Photon NOON State with Four Photons

Input photons in the state $|2\rangle_b |2\rangle_c$ ($\lambda_s = \lambda_i = 785\,\text{nm}$) were generated using the type-I pulsed-multi photon SPDC source described in Chap. 2 and launched into the chip via v-groove arrayed polarisation maintaining fibre butt-coupled to the waveguide chip.

The phase instability of states leaving outputs j and k prevents a standard tomographic approach to reconstruct the density matrix of the chip output. Ideally a reconfigurable circuit could be integrated lithographically at the output of the circuit that would be sufficient to perform full state tomography. Instead we have employed a three step process to test the coherence and measure the relative photon number of the output state of the chip: (i) temporal coherence of the multi-photon input states

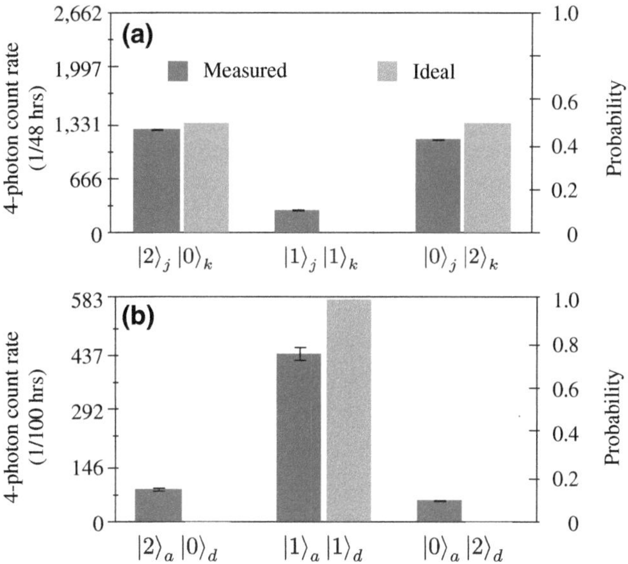

Fig. 6.2 Heralded $|2 :: 0\rangle$ state. **a** Measurement of photon statistics of the heralded two-photon NOON state. **b** Testing coherence of the heralded two-photon NOON state by measuring the photon statistics after non-classical interference at DC_2 via a fibre Sagnac loop. Both distributions are normalised using single photon detection rates to account for relative detector scheme, source and waveguide coupling efficiencies. Error bars are given the standard deviation of detected events, assuming Poissonian statistics. Figure reproduced from Ref. [1]

were verified with a generalised Hong–Ou–Mandel experiment to observe quantum interference of the state $|2\rangle_b |2\rangle_c$ incident on a 50 % reflectivity beamsplitter [17]; (ii) photon number statistics were measured at the outputs j and k (equivalent to the diagonal of the density matrix) using non-deterministic number resolving detection with optical fibre splitters; (iii) The output state was then interfered on a second beamsplitter using a directional coupler inside the chip, via an inherently phase-stable fibre Sagnac loop.

The photon number statistics measured from the heralded two-photon NOON state at outputs j and k is plotted in Fig. 6.2a. Fidelity $\left(F = \sum_j \sqrt{p_j^e p_j^m} \right)$ between the measured probability distribution (p^m) of photon statistics and the expected distribution (p^e) for the ideal state $|2 :: 0\rangle_{j,k}^0$ (also plotted) is $F_i = 0.95 \pm 0.01$. For $\lambda = 785$ nm operation, the reflectivities of DC_1 and DC_2 are measured to be $\eta = 0.542$ and 0.530 respectively. Using these measured reflectivities, the expected output state was simulated and comparison with experimental results yield a fidelity of $F_s = 0.96 \pm 0.01$, leaving the discrepancy with perfect fidelity attributed to six- and higher photon number terms from the down conversion process and residual distinguishability of photons and not the device itself.

To test the coherence of the output of the circuit we formed a Sagnac loop by joining two optical fibres coupled to modes j and k (see Sect. 6.6). This configuration results in quantum interference at DC_2 in the reverse direction and is equivalent to interference at a separate beamsplitter with zero relative optical phase of the two paths, fixed by the inherently stable Sagnac interferometer. By coupling detectors to waveguides a and d, the photon statistics of the quantum state returning through the chip after DC_2 at g and h can be measured, with an intrinsic loss due to $DC_{3,4}$. The fidelity between the measured distribution of photon statistics (Fig. 6.2b) and the distribution expected from a perfect $|2 :: 0\rangle^0_{j,k}$ state interfering at directional coupler DC_2 is $F_i = 0.90 \pm 0.03$. (Taking into account only the measured reflectivities of DC_1 and DC_2 the expected detection rates agree with the experimental measurements with fidelity $F_s = 0.97 \pm 0.03$.) Together with the temporal coherence of the input and the high fidelity of the output state in the diagonal basis, this demonstrates coherence of the $|2 :: 0\rangle_{e,f}$ state.

6.4 Heralding a Four Photon NOON State with Six Photons

Although Fig. 6.2b demonstrates coherence of the output state, the four photon process that generates it does not rely on phase stability within the interferometer structure of the device. In contrast heralding the $|4 :: 0\rangle^0_{j,k}$ state from the six photon input state $|3\rangle_b |3\rangle_c$—generated with higher-power pumping of the SPDC— requires coherent generation of the state $|3 :: 1\rangle^0_{g,h}$ within the interferometer. To test this coherence we injected the state $|3\rangle_b |3\rangle_c$ into the chip and sequentially set the phase to the four values $\phi = \pi/2, \pi, 3\pi/2, 2\pi$. On detection of the six-photon state $|1\rangle_i |4\rangle_j |0\rangle_k |1\rangle_l$, we observed the sampled interference pattern plotted in Fig. 6.3b which demonstrates twofold super-resolution (compared to the single photon interference pattern in Fig. 6.3a), and coherence of the state $|3 :: 1\rangle^0_{g,h}$ for subsequent generation of the $|4 :: 0\rangle^0_{j,k}$ state. The low number of data points does not allow fitting to a sinusoidal fringe.

Figure 6.4 shows the photon statistics of the $|4 :: 0\rangle^0_{j,k}$ state that results from the quantum interference of the state $|3 :: 1\rangle^{\pi/2}_{g,h}$ at DC_2. We fixed the phase within the chip to $\phi = \pi/2$ and again injected the six-photon state $|3\rangle_b |3\rangle_c$ into the chip. Six photons were detected in all the possible four-photon combinations on outputs j and k, together with a single photon in each of the heralding modes i and l. The fidelity between the resulting distribution of photon statistics and the distribution expected from measuring the ideal state $|4 :: 0\rangle^0_{j,k}$ is $F_i = 0.89 \pm 0.04$. Taking into account the measured reflectivities of DC_1 and DC_2, the expected statistics agree with experimental measurements with a fidelity $F_s = 0.93 \pm 0.04$. The remaining discrepancy is attributed to eight- and higher-photon number states and distinguishability of the photons generated in the down conversion process (see Sect. 6.5).

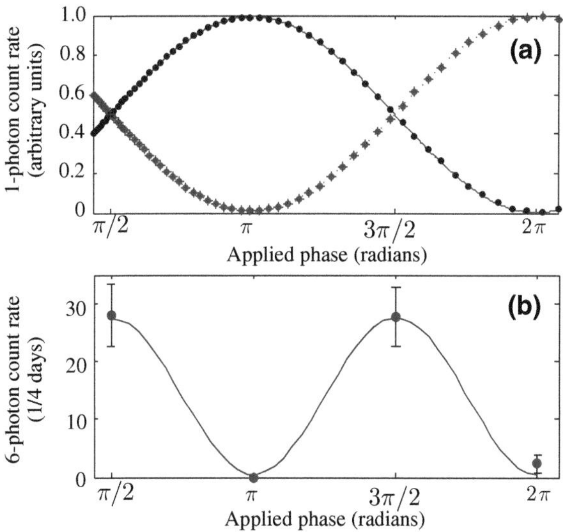

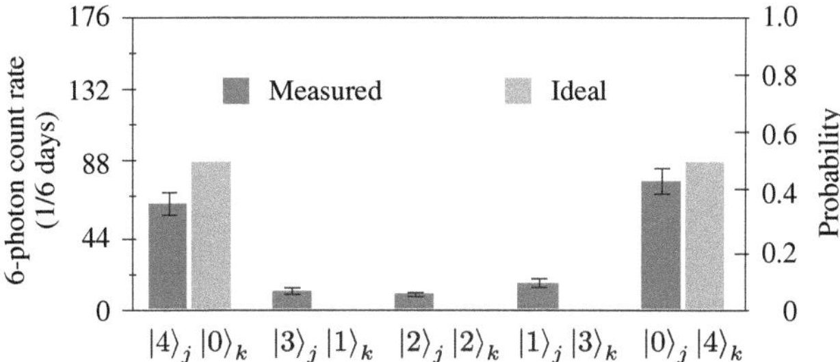

Fig. 6.3 Super resolution with a heralded four photon entangled state. **a** Single photon fringes from inputting light into waveguide b and varying the phase ϕ, displaying the expected pattern arising from classical interference pattern with period 2π. *Black circles* and *blue diamonds* respectively represent the normalised single photons count rate detected at output j and k. **b** The increased resolution interference pattern of manipulating ϕ of the state $|3 :: 1\rangle_{e,f}^{\phi}$ with period π. The four data points represent six-photon count rates integrated over four days and are normalized using single-photon count rates to account for coupling efficiency over time. Coincidence rates arising from higher photon number terms or otherwise were not subtracted from the data. *Error bars* are given the standard deviation of detected events, assuming Poissonian statistics. *Blue* sinusoidal plot of near unit contrast is plotted as a guide. Figure reproduced from Ref. [1] (colour figure online)

Fig. 6.4 Heralded generation of the $|40\rangle + |04\rangle$ state. The distribution of photon statistics from measuring a heralded four-photon NOON state. The sixfold detection rates are normalized by single photon detection rates to account for relative source, coupling and detection scheme efficiencies. Error bars are given the standard deviation of detected events, assuming Poissonian statistics. Figure reproduced from Ref. [1]

6.5 Higher Photon Number Contributions

The heralding process is tested with post-selection, using the assumption of conservation of photon number at all of the outputs of the circuit coupled to a SPCM detection scheme; this verifies that the use of number resolving detectors at outputs i and l would herald the generated entanglement without the need for counting all photons at the output of the device. Post selection also ignores lower photon number terms produced in the SPDC state (Eq. 2.5). However, the effect of higher photon number states generated in single crystal pulsed SPDC is a prominent source of noise in multi-photon experiments in general.

In the absence of number-resolving detectors in the reported experiments, the input component with eight photons $|4\rangle_b \, |4\rangle_c$ for example will give a recordable event, since losses and detectors without photon number resolution wash out the information about the input state. Non-classical interference at DC_1 transforms the eight-photon term according to

$$|4\rangle_b \, |4\rangle_c \xrightarrow{DC_1} \frac{\sqrt{35}}{8} \, |8 :: 0\rangle^0_{e,f} + \frac{\sqrt{5}}{4} \, |6 :: 2\rangle^0_{e,f} + \frac{3}{8} \, |4\rangle_e \, |4\rangle_f \qquad (6.8)$$

This can be problematic, since the eight photons can give rise to six-photon coincidental detections in different ways. For the case of $\phi = \pi/2$, the complete state that gives six-photon coincidental detection events, for example, is of the form

$$\frac{i\sqrt{2}}{162} \, |1\rangle_i \left(3\sqrt{5} \, |6 :: 0\rangle^0_{j,k} - \sqrt{3} \, |4 :: 2\rangle^0_{j,k} \right) |1\rangle_l$$

$$+ \frac{\sqrt{2}}{54\sqrt{3}} \, |2\rangle_i \left(-3\sqrt{5} \, |5 :: 0\rangle^{-\pi/2}_{j,k} - i \, |4 :: 1\rangle^{\pi/2}_{j,k} \right.$$

$$\left. + \sqrt{2} \, |3 :: 2\rangle^{-\pi/2}_{j,k} \right) |1\rangle_l$$

$$+ \frac{\sqrt{2}}{54\sqrt{3}} \, |1\rangle_i \left(3i\sqrt{5} \, |5 :: 0\rangle^{\pi/2}_{j,k} + |4 :: 1\rangle^{-\pi/2}_{j,k} \right.$$

$$\left. - i\sqrt{2} \, |3 :: 2\rangle^{\pi/2}_{j,k} \right) |2\rangle_l$$

$$+ \frac{2}{162} \, |2\rangle_i \left(-7\sqrt{3} \, |4 :: 0\rangle^0_{j,k} + 3 \, |2\rangle_j \, |2\rangle_k \right) |2\rangle_l \qquad (6.9)$$

It is clear that the eight-photon input term can give rise to a quite complex pattern of detection. In particular, all the terms in Eq. (6.9) except the NOON state terms $|N :: 0\rangle$ give rise to detection events in our setup that would be equivalent to the genuine six-photon states $|3 :: 1\rangle_{j,k}$ or $|22\rangle_{j,k}$.

To minimize the effect of the higher photon-number emission of the down-conversion process, the value of ξ was chosen to be $\xi \sim 0.085$, that corresponds to a power of the blue beam pumping the down-conversion crystal of $P_b = 215\,\mathrm{mW}$. We found this choice of ξ to be a good compromise between the integration time

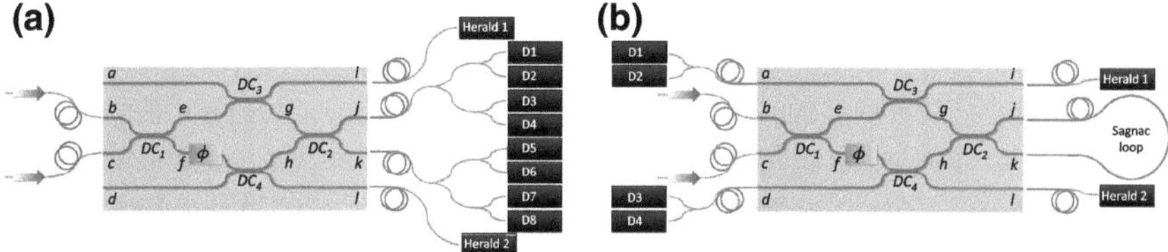

Fig. 6.5 Experimental setup for multi-photon state detection. **a** The detection scheme for detecting the six-photon state $|1\rangle_i\,|4::0\rangle_{j,k}^{\pi/2}\,|1\rangle_l$ at the output of the integrated chip. **b** The detection scheme for testing coherence of the nominally two-photon NOON state $|1\rangle_i\,|2::0\rangle_{j,k}\,|1\rangle_l$. Figure reproduced from Ref. [1]

of our six photon experiments—corresponding to a low six-photon detection rate in our experiment: $\sim$10 per day, see Fig. 6.4—and the unwanted production of eight photons. This is verified by the measured six-photon data in Fig. 6.4 where the count rates for the states $|3::1\rangle_{j,k}$ and $|22\rangle_{j,k}$ are low in comparison to the $|4::0\rangle_{j,k}$ term.

6.6 Detection Scheme

Here we describe the detection schemes used to record multi-photon states at the output of the integrated chip. The detectors used are silicon avalanche photodiode single photon counting modules (SPCM) that do not discriminate photon number–the presence of one of more photons at the SPCM produces the same output electrical signal. To reconstruct multi-photon states, number resolution is needed, which can be obtained probabilistically when using multiple SPCM and optical splitters. Three splitters and four detectors were used to detect up to four photons in the same optical mode as shown in the scheme in Fig. 6.5a. The splitters used are multi-mode fiber couplers, with a near-unity transmissivity and close to 50:50 splitting ratio. Assuming perfect 50:50 splitters, the probability of detecting four photons in the same optical mode with the above method is given by $1/4^4 \times 4! = 3/32$. Similarly, the probability of detecting three photons in one mode and one in the other is $1/4^4 \times 3! = 3/128$, and the probability of detecting two photons in two modes is $1/4^4 \times 2! = 3/256$. All multi-photon coincidental detection reported are normalized to the appropriate detection probability using single counts recorded in each detector; this normalizes the multi-photon coincidental detections, taking into account deviations from perfect and uniform detectors, differences in transmissivity and splitting ratio of the couplers.

Figure 6.5b illustrates the detection scheme used to test coherence of the nominally $|2::0\rangle_{j,k}$ state. The initial measurement of the state at the output of the chip (as drawn Fig. 6.1, yielding the data in Fig. 6.2a) proves only that the state is composed mainly of the components $|20\rangle_{j,k}$ and $|02\rangle_{j,k}$ but gives no information about the purity of the state. The coherence of the output state can be confirmed by interfering the photons in the paths j and k at a further directional coupler, since the result of this action is different in the case of a pure or a mixed state. This further interference cannot be achieved outside the integrated chip, otherwise the phase stability required for the experiment would be lost. We obtain non-classical interference at directional coupler DC_2, after the state is coupled out and back in the integrated chip via a fibre Sagnac loop between waveguides j and k, ensuring complete phase stability of the photonic state; any variation in path registered by photons traveling from waveguide j to k is experienced also by photons traveling from waveguide k to j. The photons are then probabilistically extracted from the chip via the directional couplers DC_3 and DC_4 and detected with cascaded detectors as for the other measurements.

6.7 Discussion

Heralded generation of mode-number entanglement is likely to be of use for practical application of quantum metrology. Schemes using linear optics and projective measurement have been proposed for generating large mode-number entangled states [14, 18] that scale to arbitrary size [19]. States that are robust to loss will be particularly important. The integrated waveguide architecture delivers the high stability and compact implementation required for real world applications. In particular, integrated variable beam splitters [20] will allow optimisation of quantum state engineering in presence of loss [21]. The ongoing development of efficient number resolving detectors and deterministic photon sources such as single emitters or multiplexed down-conversion schemes [22], shows promise for practical quantum metrology and other photonic quantum technologies when combined with circuits such as that described here. Real time quantum metrology requires high repetition rate (bright) sources of many photons. Future development will also require integration of fast feed-forward—using for example electro-optic materials—with the circuit demonstrated here, permitting only intended quantum metrology states to interact with measured samples, forming a building block for scalable generation of arbitrarily large entangled states [19, 23].

References

1. J.C.F. Matthews, A. Politi, D. Bonneau, J.L. O'Brien, Heralding two-photon and four-photon path entanglement on a chip. Phys. Rev. Lett. **107**, 163602 (2011)
2. H. Lee, P. Kok, N.J. Cerf, J.P. Dowling, Linear optics and projective measurements alone suffice to create large-photon-number path entanglement. Phys. Rev. A **65**, 030101 (2002)
3. E. Knill, R. Laflamme, G.J. Milburn, A scheme for efficient quantum computation with linear optics. Nature **409**(6816), 46–52 (2001)
4. J.L. O'Brien, Optical quantum computing. Science **318**(5856), 1567–1570 (2007)
5. S. Gasparoni, J.W. Pan, P. Walther, T. Rudolph, A. Zeilinger, Realization of a photonic controlled-not gate sufficient for quantum computation. Phys. Rev. Lett. **93**, 020504 (2004)
6. R. Okamoto, J.L. O'Brien, H.F. Hofmann, T. Nagata, K. Sasaki, S. Takeuchi, An entanglement filter. Science **323**(5913), 483–485 (2009)

7. X.H. Bao, T.Y. Chen, Q. Zhang, J. Yang, H. Zhang, T. Yang, J.W. Pan, Optical nondestructive controlled-not gate without using entangled photons. Phys. Rev. Lett. **98**, 170502 (2007)

8. Q. Zhang, X.H. Bao, C.Y. Lu, X.Q. Zhou, T. Yang, T. Rudolph, J.W. Pan, Demonstration of a scheme for the generation of "event-read" entangled photon pairs from a single-photon source. Phys. Rev. A **77**, 062316 (2008)

9. R. Prevedel, G. Cronenberg, M.S. Tame, M. Paternostro, P. Walther, M.S. Kim, A. Zeilinger, Experimental realization of dicke states of up to six qubits for multiparty quantum networking. Phys. Rev. Lett. **103**(2), 020503 (2009)

10. W. Wieczorek, R. Krischek, N. Kiesel, P. Michelberger, G. Toth, H. Weinfurter, Experimental entanglement of a six-photon symmetric Dicke state. Phys. Rev. Lett. **103**, 020504 (2009)

11. C. Wagenknecht, C.M. Li, A. Reingruber, X.H. Bao, A. Goebel, Y.A. Chen, Q. Zhang, K. Chen, J.W. Pan, Experimental demonstration of a heralded entanglement source. Nat. Photonics **4**, 549–552 (2010)

12. S. Barz, G. Cronenberg, A. Zeilinger, P. Walther, Heralded generation of entangled photon pairs. Nat. Photonics **4**, 553–556 (2010)

13. J. Fiurasek, Conditional generation of n-photon entangled states of light. Phys. Rev. A **65**, 053818 (2002)

14. G.J. Pryde, A.G. White, Creation of maximally entangled photon-number states using optical fibre multiports. Phys. Rev. A **68**, 052115 (2003)

15. A. Politi, M.J. Cryan, J.G. Rarity, S. Yu, J.L. O'Brien, Silica-on-silicon waveguide quantum circuits. Science **320**(5876), 646–649 (2008)

16. C.K. Hong, Z.Y. Ou, L. Mandel, Measurement of subpicosecond time intervals between two photons by interference. Phys. Rev. Lett. **59**(18), 2044–2046 (1987)

17. T. Nagata, R. Okamoto, J.L. O'brien, K. Sasaki, S. Takeuchi, Beating the standard quantum limit with four-entangled photons. Science **316**(5825), 726–729 (2007)

18. P. Kok, H. Lee, J.P. Dowling, Single-photon quantum-nondemolition detectors constructed with linear optics and projective measurements. Phys. Rev. A **66**, 063814 (2002)

19. H. Cable, J.P. Dowling, Efficient generation of large number-path entanglement using only linear optics and feed-forward. Phys. Rev. Lett. **99**, 163604 (2007)

20. J.C.F. Matthews, A. Politi, A. Stefanov, J.L. O'Brien, Manipulation of multiphoton entanglement in waveguide quantum circuits. Nat. Photonics **3**, 346–350 (2009)

21. R. Demkowicz-Dobrzanski, U. Dorner, B.J. Smith, J.S. Lundeen, W. Wasilewski, K. Banaszek, I.A. Walmsley, Quantum phase estimation with lossy interferometers. Phys. Rev. A **80**(1), 013825 (2009)

22. J.L. O'Brien, A. Furusawa, J. Vuckovic, Photonic quantum technologies. Nat. Photonics **3**, 687–695 (2009)

23. H. Cable, F. Laloe, W.J. Mullin, Noon-state formation from fock-state Bose-Einstein condensates. Phys. Rev. A **83**, 053626 (2011)

Chapter 7
Two Photon Quantum Walks

7.1 Introduction

Classical random walks are a stochastic model of a particle or walker, moving according to classical physics about a discrete space represented by a combinatoric graph—a set of vertices interconnected by edges. The application of random walks from modelling Brownian motion to constructing randomised algorithms in classical computer science [2] has inspired development of a quantum mechanical analogue in quantum computer science.

Quantum walks are a theoretical tool for quantum algorithm construction, where the quantum walker and the graph are to be simulated using a finite set of quantum gates (for example [3]). Quantum mechanics allow the walker (now a quantum particle) to adopt wave-like propagation in coherent superposition states across many positions, allowing quantum interference in the walk and therefore fundamentally different dynamics to those of a classical random walk. Potential applications include database search [4] and the graph traversal problem [5], offering respectively quadratic and exponential reduction of computation time. Furthermore, it has been proven that efficient simulation of quantum walks on graphs constructed from a particular defined set of smaller graphs (referred to as "widgets") is an alternative route to universal quantum computation [6].

As well as a tool in quantum computer science, quantum walks are also a physical phenomena, observable with an analogue approach to quantum simulation. To date, they have been realised on a number of platforms including trapped neutral atoms [7], trapped ions [8, 9], nuclear magnetic resonance [10], the frequency space of an optical resonator [11], single photons in bulk [12, 13] and fibre [14] optics and the scattering of light in coupled waveguide arrays [15]. However all such demonstrations are of a single quantum walker in a one dimensional graph. For these instances it has been shown that quantum walks have an exact mapping to classical wave phenomena

---

Some of the results reported in this chapter and an abbreviated discussion thereof were published as Ref. [1].

J. C. F. Matthews, *Multi-Photon Quantum Information Science and Technology in Integrated Optics*, Springer Theses, DOI: 10.1007/978-3-642-32870-1_7,
© Springer-Verlag Berlin Heidelberg 2013

which have an exact description using classical wave phenomena [16] as verified by experiments using bright laser light [14, 15]. Two trapped ions were used to realise a three sided coin in a quantum walk on the centre of mass of the two ions [9], but this remains a quantum walk of a single simulated particle on a discretized space of size linear in the number of resources.

This chapter reviews the model of continuous time quantum walks and in particular considers multi-photon quantum walks [17] as a generalised form of Hong–Ou–Mandel interference. We show analytically that the quantum state of indistinguishable photons on a one dimensional graph simulates a quantum walk on a nontrivial, exponentially larger graph [1]. We then present an experiment that demonstrates the continuous time quantum walk of two photons, realised in an evanescently coupled waveguide array of $N = 21$ waveguides [1].[1]

7.2 The Continuous Time Quantum Walk

Quantum walks fall under two main definitions, referred to as discrete-time and continuous-time. While the mathematical definitions are distinct, the two classes share qualitative dynamics and it has been shown that continuous time quantum walks can be reached as a limit of discrete time quantum walks [19]. We will be concerned with the class of continuous time quantum walks which is a natural model for the evolution of photons guided in evanescently coupled waveguide arrays—the experimental focus of this chapter. The definition and discussions of discrete time quantum walks can be found in appropriate reviews such as [20].

Classical random walks are typically defined to diffuse around a graph structure G which consists of a set of labelled vertices $V(G)$ (positions that can be occupied by the random walker) and edges $E(G)$ that connect vertices and define the rules for which the walker can move around G. Graphs are represented by an adjacency matrix M

$$M_{i,j} = \begin{cases} C_{i,j}, & i \neq j, \quad \text{if vertices } i \text{ and } j \text{ are connected,} \\ 0, & i \neq j, \text{ if vertices } i \text{ and } j \text{ are not connected,} \\ \beta_i, & i = j \end{cases} \quad (7.1)$$

that describe the connections between any give pair of vertices $i, j \in V$ by edges $e(i, j) \in E(G)$ and where the number of diagonal elements of M is given by the size of the graph (number of vertices). The jumping amplitude for the walker between connecting vertices is given by $C_{i,j}$ and the value β_i corresponds to the amplitude for a walker remaining at vertex i.

One of the simplest non-trivial quantum walks takes place on a uniform line-graph and is commonly used to illustrate the difference between a classical random walk

---

[1] We note that more recently, a related experiment was performed with $n = 2$ photons in a three-dimensional directly written waveguide array of $N = 6$ evanescently coupled waveguides in a ring structure [18].

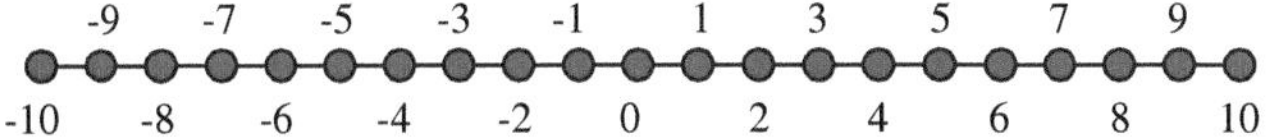

Fig. 7.1 A one dimensional graph of $N = 21$ vertices illustrating nearest-neighbour interaction

and a quantum walk. The graph shown in Fig. 7.1 consists of 21 vertices and for simplicity we assume the graph is uniform with the (sparse) tridiagonal adjacency matrix given by Eq. (7.1) for $C_{i,j} = C$ and $\beta_i = \beta$ for all i, j. The diffusive dynamics of a random walk on line-graphs lead to a Binomial distribution of detection events along the vertices of the graph, with the initial starting position as the mean [20].

Continuous time quantum walks on G are described to move around a position basis $\{|1\rangle, |2\rangle, \ldots, |N\rangle\}$ where $|i\rangle$ describes the state of a quantum mechanical walker at vertex i. The continuous time quantum walk is defined [21] with Hamiltonian $\hat{H}$ acting on the position basis according to

$$\langle i| \hat{H} |j\rangle = M_{i,j} \tag{7.2}$$

Unitary evolution of the quantum walk is therefore given by the operator

$$U(t) = e^{i\hat{H}t} \tag{7.3}$$

The probability distribution of a walker starting at a particular vertex labelled j and detected at vertex k is therefore given by

$$P = |U_{j,k}|^2 \tag{7.4}$$

For any quantum walk, there exists a time dependence of the probability distribution P given by Eq. (7.3). To illustrate the typical dynamics of quantum walks, evolution of P for the linear graph in Fig. 7.1 is given in Fig. 7.2a, b. Evolution on a larger line is given in Fig. 7.2c, d. Both studies illustrate the ballistic behaviour of the characteristic quantum walk lobes which propagate linearly in time away from the initial starting position, while the distribution left in the wake of these lobes tend to a uniform distribution with increasing time [20].

7.3 Continuous Time Quantum Walks Using Waveguide Arrays

One natural platform with which to realise continuous time quantum walks is with integrated optics using arrays of single mode, evanescently coupled waveguide (see Fig. 2.8 of Chap. 2). This approach has been applied to observe coherent evolution of continuous time quantum walk dynamics across $O(100)$ waveguides [15] and used to study experimentally the effects of Anderson localisation on quantum walks for example [22].

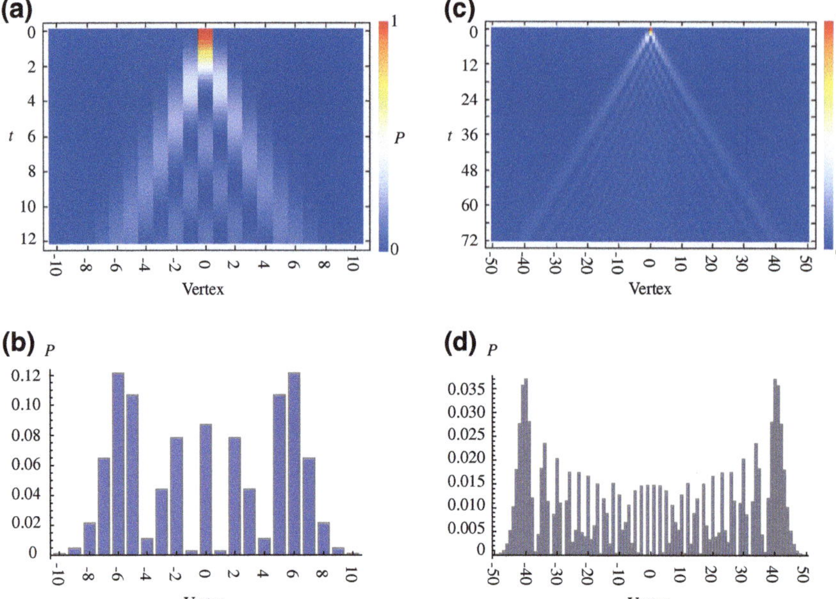

Fig. 7.2 Numerical plots of continuous time quantum walk probability distributions, illustrating the characteristic ballistic propagation the quantum walker distribution. **a** The detection probability P for finding the quantum walker at each vertex as a function of t on a 1-D graph of $N = 21$ vertices (Fig. 7.1), where the walker starts at the initial vertex labelled 0; the distribution of P at $t = 12$ is given in (**b**). **c** The detection probability P for finding the quantum walker at each vertex as a function of t on a 1-D graph of $N = 101$ vertices (Fig. 7.1), where the walker starts at the initial vertex 0; the distribution of P at $t = 72$ is given in (**d**). Distributions were computed with Eq. (7.4), for parameters $\beta = 1$, $C = 0.3$. The color-function for the probability distributions displayed in (**a**) and (**b**) are a linear scale, as shown in the corresponding legends. Simulations conducted using Mathematica

Due to the exponential decrease in the evanescent field around a single mode waveguide, this makes an ideal approximation to the nearest neighbour coupling of a quantum walk on a line. If we assume a waveguide array is uniform, then the Hamiltonian for describing waveguide array guiding single photons is given by the Hamiltonian for coupled oscillators [23], setting $\hbar = 1$:

$$\hat{H} = \sum_{j=1}^{N} \left[\beta_j a_j^\dagger a_j + C_{j,j-1} a_{j-1}^\dagger a_j + C_{j,j+1} a_{j+1}^\dagger a_j \right], \tag{7.5}$$

where the creation and annihilation operators $a_j^\dagger$ and a_j obey model excitation of a single photon in waveguide j and obey Bose–Einstein statistics.

Through choice of operator $\hat{A}$, the Heisenberg equation of motion

$$i d\hat{A}/dz = [\hat{A}, \hat{H}] \tag{7.6}$$

models the dynamics of photons propagating along distance z of the array. For example, the propagation of a single photon initially injected into waveguide k is described using $\hat{A} = a_k^\dagger$, yielding [17]

$$i\frac{\mathrm{d}a_k^\dagger}{\mathrm{d}z} = -\beta a_k^\dagger - C a_{k-1}^\dagger - C a_{k+1}^\dagger.$$

(7.7)

Operating Eq. (7.7) on the vacuum state $|0\rangle$, and inspecting in the Schrödinger picture we $i\mathrm{d}|\psi_1\rangle/\mathrm{d}z = H^{(1)}|\psi_1\rangle$ for $|\psi_1\rangle$ restricted to the single photon Fock state space. This leads to the effective Hamiltonian $H^{(1)}$ taking the form

$$H^{(1)} = \sum_k \beta\,|1\rangle_k{}_k\langle 1| + C\,|1\rangle_{k-1}{}_k\langle 1| + C\,|1\rangle_{k+1}{}_k\langle 1|$$

(7.8)

from which the adjacency matrix $M_{j,k}^{(1)} = {}_j\langle 1|\,H^{(1)}\,|1\rangle_k$ of the graph in Fig. 7.1 is constructed using representation in the single photon Fock basis $\{|1\rangle_i = a_i^\dagger|0\rangle\}$. This is equivalent to the adjacency matrix and Hamiltonian given in Eqs. (7.1) and (7.2). It therefore follows that the unitary evolution operator equivalent to a uniform waveguide array of finite length z, assuming no loss, is modelled by

$$U = e^{iH^{(1)}z/c}$$

(7.9)

where the propagation distance along the waveguide $z = ct$ replaces the time parameter in Eq. (7.3).

7.4 Correlated Quantum Walk of Two Photons

A model of two entangled quantum walks was described theoretically in 2006 by Omar et al. [24] and was based upon entangling the coins of two discrete time quantum walkers on two separate graphs. In 2009, Bromberg et al. [17] proposed a model of two indistinguishable photons undergoing a generalisation of Hong–Ou–Mandel interference in a continuous time quantum walk unitary that would be realised in a waveguide array. More recently, similar experiments have been proposed to study correlations in the presence of Anderson localisation [25] and Bloch oscillations of two photon NOON states in waveguide arrays [26] .

The underlying idea of a two photon continuous time quantum walk is to observe two photon correlations, arising from quantum interference in a continuous time quantum walk unitary. In the Heisenberg picture, single photon evolution is given by the mode transformation

$$a_j^\dagger|0\rangle \xrightarrow{U} \sum_l U_{l,j}a_l^\dagger|0\rangle$$

(7.10)

which yields the distribution

$$| \langle 0| a_q \sum_l U_{l,j} a_l^\dagger |0\rangle |^2 = |U_{l,j}|^2 \tag{7.11}$$

for detecting a single photon at waveguide output labeled q.

Similarly, two photon input into waveguides j and k evolve according to the mode transformation

$$a_j^\dagger a_k^\dagger |0\rangle \xrightarrow{U} \left(\sum_l U_{l,j} a_l^\dagger \right)\left(\sum_m U_{m,k} a_m^\dagger \right) |0\rangle \tag{7.12}$$

The correlated detection pattern of two photons at output s and t is therefore given by a matrix of correlations

$$\Gamma_{q,r} = |U_{q,j} U_{r,k} + U_{q,k} U_{r,j}|^2 \tag{7.13}$$

which is derived by computing the inner product

$$\langle 0| a_q a_r \left(\sum_l U_{l,j} a_l^\dagger \right)\left(\sum_m U_{m,k} a_m^\dagger \right) |0\rangle \tag{7.14}$$

using the Bose–Einstein commutation relations (a more detailed treatment follows in Chap. 8). For normalisation of Fock states of more than one photon—$a^{\dagger n} |0\rangle = \sqrt{n!} |n\rangle$—the probability to detect two photons at the same outputs[2] q is given by $P_{q,q} = \frac{1}{2}\Gamma_{q,q}$: for all other $q \neq r$ we have $P_{q,r} = \Gamma_{q,r}$ [27]. For distinguishable photons launched into inputs j, k, the quantum walk dynamics are given by

$$P_{q,r} = |U_{s,j} U_{t,k}|^2 + |U_{t,j} U_{s,k}|^2 \tag{7.15}$$

Using Eq. (7.13), simulations are straight forward for a variety of input states and a range of time evolutions t of the operator $U = e^{i\hat{H}t}$. Figure 7.3 displays simulations of $\Gamma_{q,r}$ for two boson correlated quantum walks on a uniform and linear graph of $N = 101$ for the four different input states $a_0^\dagger a_1^\dagger |0\rangle$, $a_{-1}^\dagger a_1^\dagger |0\rangle$, $a_{-1}^\dagger a_2^\dagger |0\rangle$ and $a_{-2}^\dagger a_2^\dagger |0\rangle$. Four correlation matrices are simulated for three values of t for each of the example input states. Studies (a)–(f) are similar to those of [17].

We first note that through the correlations of quantum mechanics, detection of one particle effects the observed dynamics of the second particle in the quantum walk. Detecting one photon in a two photon quantum walk can therefore be considered to be a form of heralded quantum walk. The second point to notice is that, although our study is far from exhaustive, the examples illustrate that different input

---

[2] The $\frac{1}{2}$ factor for the diagonal elements arises from the normalisation of the Fock state $a_j^{\dagger 2} |0\rangle = \sqrt{2!} |2\rangle$.

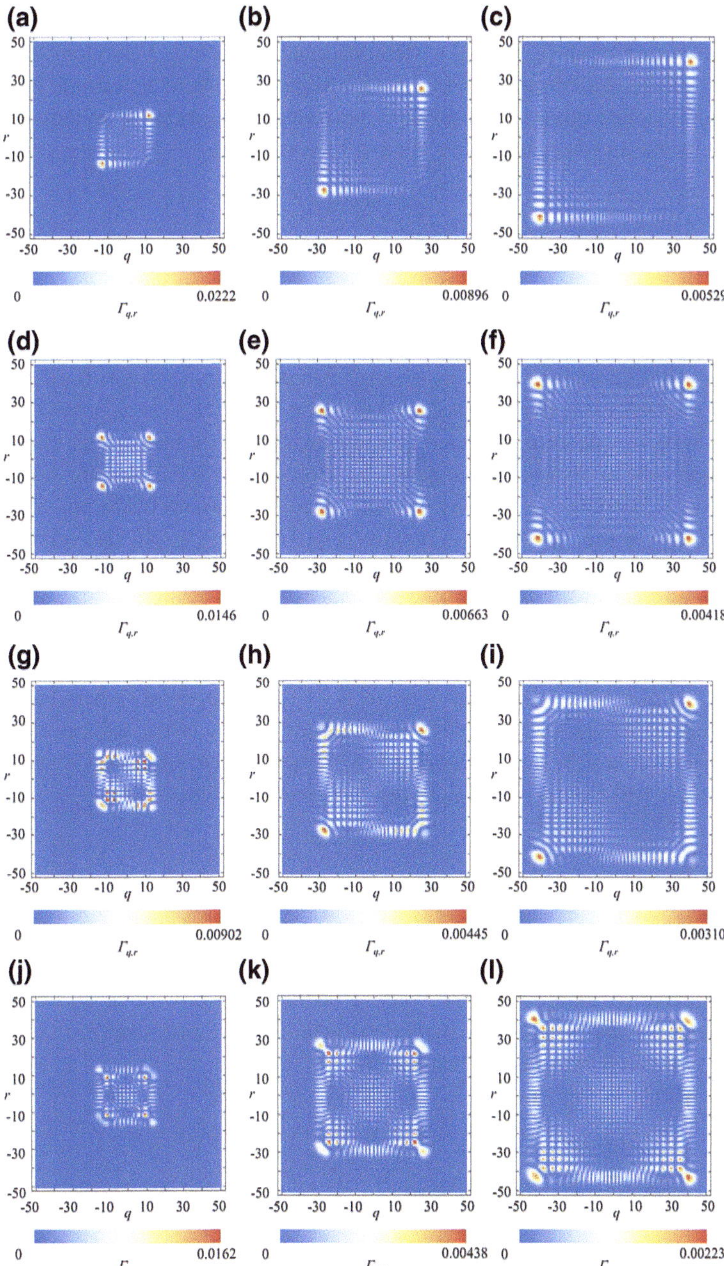

Fig. 7.3 Mathematica simulations of two photon correlated quantum walks on a 101 vertex linear and uniform graph, as described by the correlation matrices $\gamma_{q,r}$ for given in Eq. (7.13). Graph parameters are $\beta = 1$, $C = 0.3$ and the simulated inputs are: $j = 0$, $k = 1$ (**a**–**c**); $j = -1$, $k = 1$ (**d**–**f**); $j = -1$, $k = 2$ (**g**–**i**); and $j = -2$, $k = 2$ (**j**–**l**). Three time evolutions are presented for each input: $t = 24$ (**a, d, g, j**); $t = 48$ (**b, e, h, k**); and $t = 72$ (**c, f, i, l**)

states effect the overall dynamics of the two photon quantum walks. For example for the neighbouring input (Fig. 7.3a–c) demonstrates a bunching behaviour that would appear similar to that observed in Hong–Ou–Mandel interfere on a single beamsplitter. However, while other inputs do show boson bunching behaviour, they also have the capacity for anti-bunching (Fig. 7.3d–f, for example).

7.5 Simulating Quantum Walks Using Photonic Quantum Interference

A single quantum walker evolving around a labelled, finite graph G of size N is confined to a Hilbert space spanned by a position basis $\{|1\rangle, |2\rangle, \ldots, |N\rangle\}$. Suppose now we have m quantum particles moving around G. A new graph can then be constructed whose vertices correspond for example to the spanning set of Fock states of the m particles, occupying a subset of the N positions of the physical graph. These states will then evolve, in a manner dependent on the physical graph G, but equivalent to the evolution of a single particle moving on a new graph. In this sense a multi-particle quantum walk on G simulates the quantum walk of a single particle moving around a larger graph of size $O(N^m/m!)$ [1].[3]

The graph illustrated in Fig. 7.4a is associated to one photon undergoing a continuous time quantum walk in an array of $N = 21$ waveguides, where each vertex corresponds to a Fock state of one photon occupying one of the waveguides. Using the coupled oscillator Hamiltonian $\hat{H}$ given in Eq. (7.5) and applying the Heisenberg equation Eq. (7.6) for the two photon input $\hat{A} = a_j^\dagger a_k^\dagger$ yields

$$i\frac{\mathrm{d}a_j^\dagger a_k^\dagger}{\mathrm{d}z} = -2\beta a_j^\dagger a_k^\dagger - C\left[a_j^\dagger a_{k-1}^\dagger + a_j^\dagger a_{k+1}^\dagger + a_k^\dagger a_{j-1}^\dagger + a_k^\dagger a_{j+1}^\dagger\right] \qquad (7.16)$$

Operating Eq. (7.16) on the vacuum state $|0\rangle$ and inspecting in the Schrödinger picture reveals an effective Hamiltonian $H^{(2)}$ acting on the two photon Fock state space from which an adjacency matrix $M^{(2)}$ can be extracted by first assigning each element to a pair of two given two-photon Fock states;

$$|j'\rangle = |1\rangle_j |1\rangle_k \,;\, |s'\rangle = |1\rangle_s |1\rangle_t \qquad (7.17)$$

and then computing the inner product

$$M^{(2)}_{j',s'} = \langle j'| H^{(2)} |s'\rangle \qquad (7.18)$$

---

[3] We note that the simulation of higher dimensional discrete time quantum walks using two photons is treated in [28].

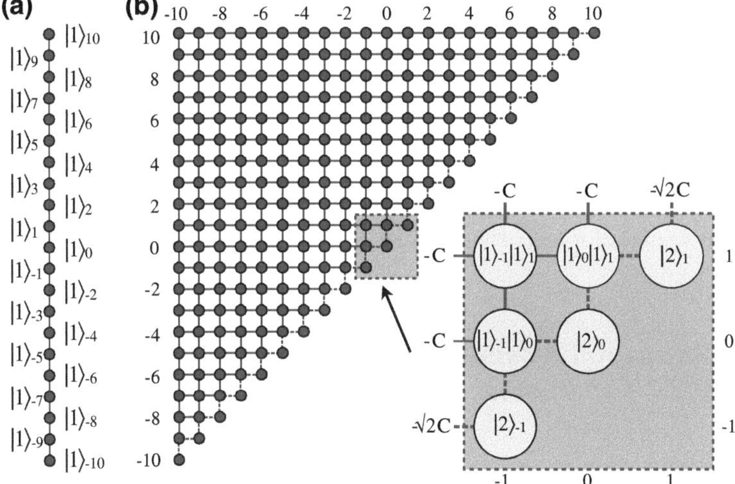

Fig. 7.4 One and two dimensional lattice graphs. **a** The one dimensional graph of vertices representing the state space of one photon populating $N = 21$ a one-dimensional array of evanescently coupled waveguides with nearest neighbour coupling. Site potentials for each vertex are $-\beta$ while hopping amplitudes between nearest neighbours are all equal to $-C$. **b** The two-dimensional lattice of vertices that represent the two-photon state space populating $N = 21$ waveguides in a coupled one-dimensional waveguide array (*inset*). The enlarged region of the lattice displays the vertex representation of the two photon basis states. Site potentials are all equal to -2β, while hopping amplitudes between adjacent vertices are either $-C$ or $-\sqrt{2}C$ as labeled. Figure reproduced from Ref. [1]

for all two photon Fock states. This adjacency matrix, for a uniform waveguide array described by Eq. (7.5), is equivalent to the two dimensional lattice given in Fig. 7.4b.

The indistinguishability of photons (or other particles) used for such a physical simulation is important to realise the perpendicular lattice structure shown in Fig. 7.4. In general, the Hilbert space representing n distinguishable photons injected into a waveguide array of size N is the tensor product of n Hilbert spaces each of dimension N. This makes the combined state space N^n dimensional. However, the photons still remain in a separable product state throughout evolution with unitary transform in the form of a tensor product of n single photon unitary operators. These dynamics are trivial to reconstruct, requiring for example n repetitions of the single photon evolution experiment (equivalently bright laser light). The resulting graph would be equivalent to a cartesian product of one dimensional graphs: For $n = 2$, this would contain the general structure of the lattice in Fig. 7.4 as a subgraph, but would also contain diagonal edges that connect directly edges (j, k) and $(j + 1, k + 1)$ for example. The dynamics of a continuous time quantum walk on a cartesian product of two graphs G_1 and G_2 could be observed by combining the results of two individual bright light experiments on G_1 and G_2 performed separately.

For n indistinguishable photons, the dimension of the Hilbert space grows according to $O(N^n/n!)$. However, unlike the distinguishable case, evolution moves into non-separable states via quantum interference—for example, number-mode entangled NOON states (see Chaps. 3, 5 and 6). For the case of two indistinguishable photons ($\hat{A} = a_j^\dagger a_k^\dagger$) injected to array of size N, the Hilbert space is of dimension $(N + 1)N/2$ (proven by induction). For the further example of 3-photons ($\hat{A} = a_j^\dagger a_k^\dagger a_l^\dagger$), the Hilbert space is of dimension $(N + 2)(N + 1)N/6!$ (proven by induction); this emulates a CTQW on a three dimensional lattice. In general, for a quantum walk of n indistinguishable photons, the dynamics cannot be decomposed into the dynamics of n independent walks.

7.6 Experimental Realisation of a Correlated Quantum Walk of Two Photons

Degenerate photon pairs were produced via CW pumped, type I spontaneous parametric down conversion (see Chap. 2) and launched into a 750 μm long Silicon Oxynitride waveguide array, as discussed in Sect. 2.4.3. The two central, two-photon input states that provide maximum overlap in the waveguide array were measured using the input states $a_0^\dagger a_1^\dagger$ and $a_{-1}^\dagger a_1^\dagger$. Multiple commercially available avalanche photodiode single photon counting modules were used in conjunction with FPGA coincidence counting logic to record two photon coincidence events at the output of the array.

The measured correlated photon detection rates at the output of the array are plotted in Fig. 7.5 for launching photon pairs into inputs 0 and 1. The correlations are plotted as matrices, with elements $P_{q,r}$ given by the outcome of detecting a two photon coincidence event at outputs labelled q and r. Two sets of data were collected: the first (Fig. 7.5b) uses a temporal delay outside of the coherence length of the photons to make the photons distinguishable [29] (not overlapped); the second set (Fig. 7.5d) is of indistinguishable photon pairs launched with zero temporal into the array. The overlap of these measured distributions with ideal simulations (Fig. 7.5a, c) are $S = 0.980 \pm 0.001$ and $S = 0.934 \pm 0.001$, respectively, where S is the similarity between two probability distributions P and P' defined by

$$S = \left(\sum_{i,j} \sqrt{P_{i,j} P'_{i,j}} \right)^2 \bigg/ \sum_{i,j} P_{i,j} \sum_{i,j} P'_{i,j} \qquad (7.19)$$

The smaller value of S in the overlapped case is attributed to partial distinguishability due to imperfections in the photon source, leading to non-perfect quantum

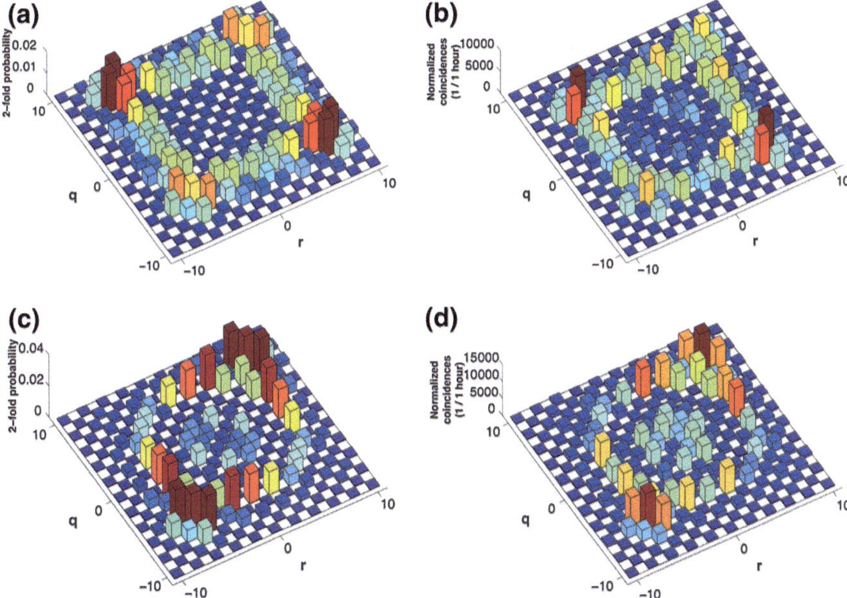

Fig. 7.5 Two photon coincidence matrices for photon pairs launched into inputs ($q = 0$, $r = 1$). **a** Theoretical plot of $P_{q,r}$ for distinguishable photons. **b** Measured correlated detection statistics of two distinguishable photons at the output of the waveguide array. **c** Theoretical plot of $P_{q,r}$ for indistinguishable photons. **d** Measured correlated detection statistics of two indistinguishable photons at the output of the waveguide array. Figure from Ref. [1]

interference. Characteristic boson bunching of the photons can be observed in this data, in the diagonal elements of the matrix that correspond to the two photons detected in the same waveguide. The vanishing of the two off-diagonal lobes is a result of the destructive interference of quantum amplitudes in the correlation matrix and can be considered as a generalisation of the Hong–Ou–Mandel experiment.

A different quantum interference image is observed on injecting two photons in two waveguides -1 and 1 with initial state $a^{\dagger}_{-1} a^{\dagger}_{1} |0\rangle$. The measured data are plotted in Fig. 7.6 for the case of distinguishable photons (ideal theoretical Fig. 7.6a and experimental Fig. 7.6b distributions) and indistinguishable photons (ideal theoretical Fig. 7.6c and experimental Fig. 7.6d distributions). The similarities with the ideal simulation are $S = 0.970 \pm 0.002$ and $S = 0.903 \pm 0.002$ for the distinguishable and indistinguishable photons respectively. In this second case, both bunching and anti-bunching is observed with the main feature being the vanishing of probability to simultaneously detect one photon in the centre of the array and to detect one at the limit of ballistic propagation (for example in waveguides $q = 7$ and $r = 0$ in Fig. 7.6c, d).

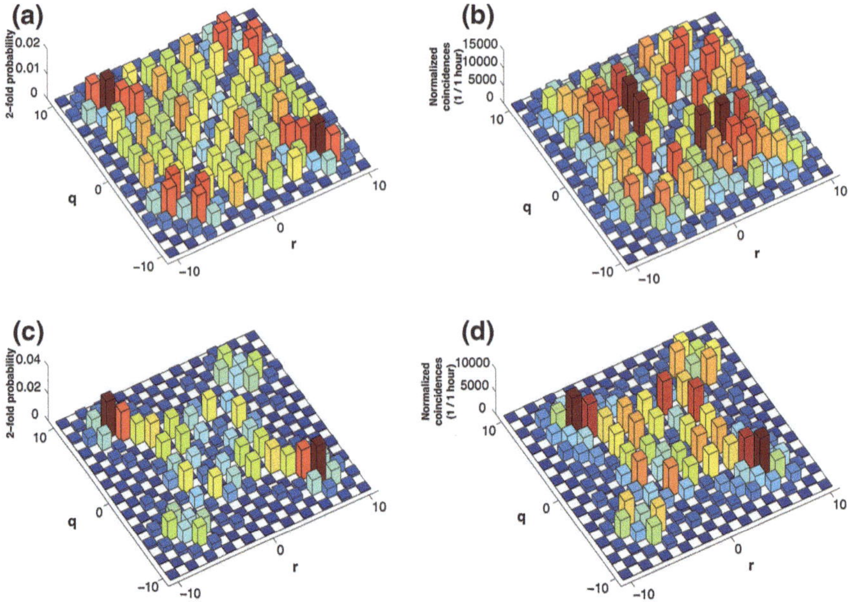

Fig. 7.6 Two photon coincidence matrices for photon pairs launched into inputs $(-1, 1)$. **a** Theoretical plot of $P_{q,r}$ for distinguishable photons. **b** Measured correlated detection statistics of two distinguishable photons at the output of the waveguide array. **c** Theoretical plot of $P_{q,r}$ for indistinguishable photons. **d** Measured correlated detection statistics of two indistinguishable photons at the output of the waveguide array. Figure from Ref. [1]

7.7 Discussion

We have seen that multi-photon quantum walks can be used to simulate single particle quantum walks on more complex structures; more explicitly, increasing the photon number n emulate quantum walks on hyper-cubic graphs exponentially large in n. Combining this with quantum interference in truly three-dimensional directly written waveguides [18, 30] would allow further increase in graph size and complexity. Furthermore, three dimensional, directly written implementations of two dimensional graphs, with classical laser light as input, would verify the analytical model that two photons injected into a one-dimensional graph would simulate one particle undergoing a quantum walk in two dimensions. A natural and immediate experimental extension to the two photon quantum walks reported here, would be to increase the number of particles used.

In our theoretical and experimental studies, we have modelled and used arrays of waveguide with fixed propagation and coupling coefficients $\beta_j = \beta$ and $C_{j\pm1} = C$; independent variation these parameters for each j provides a means to engineer graph structure and therefore the dynamics of quantum walks. For example, varying these parameters randomly and independently allows investigation of correlated quantum walks in disordered systems and observe the effects of Anderson localisation, known to affect propagation of quantum information [22, 31]. The ability to tailor these parameters will also allow studies into varied propagation dynamics of the quantum information flowing around a quantum walk, including perfect state transfer [32, 33].[4]

Another key aspect is precisely how quantum particle statistics effect the dynamics of multi-particle quantum walks. This may have applications for modelling multiple excitations in quantum walks, or multi-particle simulations of graphs with specific properties (e.g. [35]). In this chapter, we are restricted to Bose–Einstein statistics since we are considering and using photons only. In the next chapter, we provide a means to simulate different quantum statistics using entangled photons.

References

1. A. Peruzzo, M. Lobino, J.C.F. Matthews, N. Matsuda, A. Politi, K. Poulios, X.-Q. Zhou, Y. Lahini, N. Ismail, K. Wörhoff, Y. Bromberg, Y. Silberberg, M.G. Thompson, J.L. O'Brien, Quantum walks of correlated photons. Science **329**, 1500–1503 (2010)
2. R. Motwani, P. Raghavan, *Randomized Algorithms* (Cambridge University Press, New York, 1995)
3. A. Ambainis, Quantum walk algorithm for element distinctness. SIAM J. Comput. **37**, 210–239 (2007)
4. A.M. Childs, J. Goldstone, Spatial search by quantum walk. Phys. Rev. A **70**, 022314 (2004)
5. A.M. Childs, R. Cleve, E. Deotto, E. Farhi, S. Gutmann, D.A. Spielman, Exponential algorithmic speedup by a quantum walk, in *STOC '03: Proceedings of the Thirty-fifth Annual ACM Symposium on Theory of Computing*, 2003, pp. 59–68
6. A.M. Childs, Universal computation by quantum walk. Phys. Rev. Lett. **102**, 180501 (2008)
7. M. Karski, L. Forster, J.-M. Choi, A. Steffen, W. Alt, D. Meschede, A. Widera, Quantum walk in position space with single optically trapped atoms. Science **325**(5937), 174–177 (2009)
8. H. Schmitz, R. Matjeschk, C. Schneider, J. Glueckert, M. Enderlein, T. Huber, T. Schaetz, Quantum walk of a trapped ion in phase space. Phys. Rev. Lett. **103**, 090504 (2009)
9. F. Zähringer, G. Kirchmair, R. Gerritsma, E. Solano, R. Blatt, C.F. Roos, Realization of a quantum walk with one and two trapped ions. Phys. Rev. Lett. **104**, 100503 (2010)
10. C.A. Ryan, M. Laforest, J.C. Boilequ, R. Laflamme, Experimental implemtation of a discrete-time quanutm random walk on an nmr quantum-information processor. Phys. Rev. A **72**, 062317 (2005)
11. D. Bouwmeester, I. Marzoli, G.P. Karman, W. Schleich, J.P. Woerdman, Optical galton board. Phys. Rev. A **61**, 013410 (1999)
12. B. Do, M. Stohler, S. Balasubramanian, D. Elliott, Experimental realization of a quantum quincunx by use of linear optical elements. JOSA B **22**(2), 499–504 (2005)
13. M.A. Broome, A. Fedrizzi, B.P. Lanyon, I. Kassal, A. Aspuru-Guzik, A.G. White, Discrete single-photon quanutm walks with tunable decoherence. Phys. Rev. Lett. **104**, 153602 (2010)
14. A. Schreiber, K.N. Cassemiro, V. Potoček, A. Gábris, P.J. Mosley, E. Andersson, I. Jex, C. Silberhorn, Photons walking the line: a quantum walk with adjustable coin operations. Phys. Rev. Lett. **104**(5), 050502 (2010)
15. H.B. Perets, Y. Lahini, F. Pozzi, M. Sorel, R. Morandotti, Y. Silberberg, Realization of quantum walks with negligible decoherence in waveguide lattices. Phys. Rev. Lett. **100**(17), 170506 (2008)

---

[4] In for these two references, the quantum walk is modelled as a spin chain for transferring quantum information [34].

16. P.L. Knight, E. Roldán, J.E. Sipe, Quantum walk on the line as an interference phenomenon. Phys. Rev. A **68** 020301(R) (2003)
17. Y. Bromberg, Y. Lahini, R. Morandotti, Y. Silberberg, Quantum and classical correlations in waveguide lattices. Phys. Rev. Lett. **102**(25), 253904 (2009)
18. J.O. Owens, M.A. Broome, D.N. Biggerstaff, M.E. Coggin, A. Fedrizzi, T. Linjordet, M. Ams, G.D. Marshall, J. Twamley, M.J. Withford, A.G. White, Two-photon quantum walks in an elliptical direct-write waveguide array. New J. Phys. **13**, 075003 (2011)
19. A.M. Childs, On the relationship between continuous- and discrete-time quantum walk. Commun. Math. Phys. **294**, 581–603 (2010)
20. J. Kempe, Quantum random walks: an introductory overview. Contemp. Phys. **44**(4), 307–327 (2003)
21. A.M. Childs, E. Farhi, S. Gutmann, An example of the difference between quantum and classical random walks. Quantum Inf. Process. **1**, 35–43 (2002)
22. Y. Lahini, A. Avidan, F. Pozzi, M. Sorel, R. Morandotti, D.N. Christodoulides, Y. Silberberg, Anderson localization and nonlinearity in one-dimensional disordered photonics lattices. Phys. Rev. Lett. **100**, 013906 (2008)
23. L.E. Estes, T.H. Keil, L.M. Narducci, Quantum-mechanical description of two coupled harmonic oscillators. Phys. Rev. **175**(1), 286 (1968)
24. Y. Omar, N. Paunković, L. Sheridan, S. Bose, Quantum walk on a line with two entangled particles. Phys. Rev. A **74**, 042304 (2006)
25. Y. Lahini, Y. Bromberg, D.N. Christodoulides, Y. Silberberg, Quantum correlations in two-particle anderson localization. Phys. Rev. Lett. **105**, 163905 (2010)
26. Y. Bromberg, Y. Lahini, Y. Silberberg, Bloch oscillations of path-entangled photons. Phys. Rev. Lett. **105**, 263604 (2010)
27. K. Mattle, M. Michler, H. Weinfurter, A. Zeilinger, M. Zukowski, Non-classical statistics at multiport beam splitter. Appl. Phys. B: Lasers Opt. **60**(2–3, Suppl. S), S111 (1995)
28. P. Rohde, A. Schreiber, M. Stefanak, I. Jex, C. Silberhorn, Multi-walker discrete time quantum walks on arbitrary graphs, their properties, and their photonic implementation. New J. Phys. **13**, 013001 (2011)
29. C.K. Hong, Z.Y. Ou, L. Mandel, Measurement of subpicosecond time intervals between two photons by interference. Phys. Rev. Lett. **59**(18), 2044–2046 (1987)
30. E. Keil, A. Szameit, F. Dreisow, M. Heinrich, S. Nolte, A. Tünnermann, Photon correlations in two-dimensional waveguide arrays and their classical estimate. Phys. Rev. A **81**, 023834 (2010)
31. J.P. Keating, N. Linden, J.C.F. Matthews, A. Winter, Localization and its consequences for quantum walk algorithms and quantum communication. Phys. Rev. A **76**, 012315 (2007)
32. M. Christandl, N. Datta, A. Ekert, A.J. Landahl, Perfect state transfer in quantum spin networks. Phys. Rev. Lett. **92**, 187902 (2004)
33. M.-H. Yung, Quantum speed limit for perfect state transfer in one dimension. Phys. Rev. A **74**, 030303(R) (2006)
34. S. Bose, Quanutm communication through an unmodulated spin chain. Phys. Rev. Lett. **91**, 207901 (2003)
35. J.M. Harrison, J.P. Keating, J.M. Robbins, Quantum statistics on graphs. Proc. R. Soc. A **467**, 212–233 (2011)

Chapter 8
Simulating Arbitrary Quantum Statistics with Entangled Photons

8.1 Introduction

Photons are classed as bosons with corresponding non-classical interference behaviour accurately modelled by Bose–Einstein statistics. The low noise properties of photons make them ideal for exploring quantum phenomena. However, photons can be viewed as limited in their use, due to their lack of interaction and their restriction to behaviour dictated by Bose–Einstein statistics, which has implications for quantum interference [2] and quantum logic gates [3], for example. Here we show that by controlling one phase parameter, entanglement can exactly simulate quantum dynamics in an arbitrary mode transformation—labelled A—of N identical fermions, bosons or particles with fractional statistics. We describe a generalised approach to simulate quantum interference of many particles with Fermi–Dirac statistics and a form of fractional statistics residing in the intermediate regime between Bose–Einstein and Fermi–Dirac statistics. Finally we report an experiment with two polarisation entangled photons to simulate the quantum interference of two Bosons, two Fermions and in general two particles with fractional statistics, undergoing quantum interference in a continuous time quantum walk unitary.

The ability to simulate non-bosonic statistics with photons could give access to phenomena which are not otherwise physically accessible or would be obscured by decoherence; provide a means to verify quantum simulations performed in other quantum systems [4, 5]; and overcome the difficulty of separating external phase contributions from genuine anyon statistics [6]. Applications include simulating coupled spin systems such as arrays of quantum dots [7, 8] and atomic ensembles in optical lattices [9]. In general, computation of correlations that result from multi-particle, multi-path quantum interference is not efficient [10]. However, as we saw in the previous chapter, large scale repeated quantum interference is relatively straightforward to achieve with photons in waveguides [11].

Some of the results reported in this chapter and an abbreviated discussion thereof are in the submitted Ref. [1].

J. C. F. Matthews, *Multi-Photon Quantum Information Science and Technology in Integrated Optics*, Springer Theses, DOI: 10.1007/978-3-642-32870-1_8,
© Springer-Verlag Berlin Heidelberg 2013

8.2 Quantum Particle Statistics and Correlation Functions

In quantum mechanics, fundamental particle classes are characterised by their behaviour under pairwise exchange ($a_j^\dagger a_k^\dagger = e^{i\phi} a_k^\dagger a_j^\dagger$). For example an arbitrary number of bosons (obeying Bose–Einstein statistics: $\phi = 0$) can occupy a single quantum state, leading to bunching and Bose–Einstein condensation, whereas fermions (obeying Fermi–Dirac statistics: $\phi = \pi$) are strictly forbidden to occupy the same state via the Pauli-exclusion principle [12, 13]. In two dimensions particles known as anyons [14] can exhibit intermediate behaviour with fractional exchange statistics a concept applied in superconductivity [15] and used to explain the fractional quantum Hall effect [16].

Exchange statistics refer to a global wave-function acquiring phase ϕ after interchanging two indistinguishable quantum particles according to the relations

$$a_j^\dagger a_k^\dagger - e^{i\phi} a_k^\dagger a_j^\dagger = 0 \tag{8.1}$$

$$a_j a_k - e^{i\phi} a_k a_j = 0 \tag{8.2}$$

$$a_j a_k^\dagger - e^{i\phi} a_k^\dagger a_j = \delta_{j,k} \tag{8.3}$$

for $j < k$. In three dimensions particles are either bosons ($\phi = 0$) or fermions ($\phi = \pi$) and successive swapping is represented by the permutation group while the correlated detection statistics are associated respectively to the permanent and determinant of a given matrix [2]. In one- and two-dimensions particles referred to as anyons [14] can exist where swapping two identical anyons (represented by the braid group) results in the combined wave-function acquiring phase $\phi = \theta \omega$ where $0 < \theta < \pi$ and the positive (or negative) winding number ω keeps track of the number of times two anyons braid in trajectory space with clockwise (or anti-clockwise) orientation [15].

The relations Eqs. (8.1)–(8.3) directly impart a ϕ dependence on the correlated detection statistics of indistinguishable particles undergoing quantum interference. Chapter 7 considered correlation functions for the particular case $\phi = 0$.

8.2.1 Deriving the Two-Particle Correlation Function for Arbitrary Statistics

Suppose we have a process A represented by a matrix whose element $A_{s,j}$ is the complex transition amplitude from j to s. Transformation A on a state of two indistinguishable particles created in modes j and k, therefore maps the initial state $a_j^\dagger a_k^\dagger |0\rangle$ to a linear superposition of two particle states according to

$$a_j^\dagger a_k^\dagger |0\rangle \xrightarrow{A} \sum_{p,q} A_{p,j} A_{q,k} a_p^\dagger a_q^\dagger |0\rangle , \tag{8.4}$$

where A is expressed as a matrix in the mode basis. Correlated detection at s and t will occur as a result of two possible events: (i) the particle at j is mapped to position s and the particle at k is mapped to t with amplitude $A_{j,s}A_{k,t}$; (ii) the particle at j is mapped to position t and the particle at k is mapped to s with amplitude $A_{j,t}A_{k,s}$. Labelling modes such that $j < k$ and $s < t$ implies (ii) is equivalent to at least one exchange of the position of the two particles. Here we define (i) corresponds to no exchange while (ii) corresponds to a single swap or braid operation according to (8.1–8.3).

Since the two particles are indistinguishable, quantum mechanics dictates outcomes (i) and (ii) are indistinguishable and hence the complex probability amplitudes $A_{j,s}A_{k,t}$ and $A_{j,t}A_{k,s}$ interfere in the correlation across s and t. The correlated detection outcome $\Gamma^{\phi}_{s,t}$ is therefore computed by projecting the state $\langle 0| a_s a_t$ on the output state (8.4), using the inner product

$$\langle 0| a_s a_t a_j^{\dagger} a_k^{\dagger} |0\rangle = \delta_{j,t}\delta_{k,s} + e^{i\phi}\delta_{j,s}\delta_{k,t}, \tag{8.5}$$

(which follows from the relations (8.1–8.3)). Combining expressions (8.4) and (8.5) therefore yields the ϕ dependent correlation function

$$\Gamma^{\phi}_{s,t} = \left| A_{s,j}A_{t,k} + e^{i\phi}A_{t,j}A_{s,k} \right|^2 \tag{8.6}$$

for exchange statistics parameterised by ϕ.

Non-classical behaviour is evident in the constructive and destructive interference between typically complex terms inside the $|\cdot|^2$, dictated by ϕ. In order to observe the effect of fractional exchange statistics, we define a situation that allows quantum interference between two indistinguishable anyons undergoing braiding operations dependent upon A; expression (8.6) represents the resulting quantum interference between two abelian anyons ($0 < \phi < \pi$) in superposition of braiding once and not braiding at all. To see this we define $j < k$ and $s < t$ and consider two identical anyons created at modes j and k and transformed according to A. Correlated detection at outputs s and t reveals two indistinguishable possibilities that correspond to the two terms in Eq. (8.6): A maps the anyon at j to mode s and the anyon at k to mode t with amplitude $A_{s,j}A_{t,k}$; or for A to map the anyon at j to mode t and the anyon at k to mode s via a clockwise braid with fractional phase ϕ and amplitude $A_{t,j}A_{s,k}$. From this definition, we can use relations (8.1–8.3) to derive expression (8.6) for $0 < \phi < \pi$.

The quantity $\Gamma^{\phi}_{s,t}$ been derived for photons (bosons) [17] and fermions [2] in multi-port networks. Note that for fermions ($\phi = \pi$) the diagonal elements of the matrix $\Gamma^{\pi}_{s,s} = 0$ agree with the Pauli-exclusion principle.

8.2.2 The N-Particle Correlation Function
for Arbitrary Statistics

Consider N particles with exchange parameter ϕ and governed by the relations (8.1–8.3) which are launched into inputs labelled by tuple $\vec{\nu} = \{\nu_1, \nu_2, \ldots, \nu_N\}$ of mode transformation U. The N-fold correlation function for detection across output N-tuple $\vec{\mu} = \{\mu_1, \mu_2, \ldots, \mu_N\}$ is derived for multi-particle quantum walks with both Fermi–Dirac and Bose–Einstein statistics [18], which naturally extends to arbitrary statistics and an arbitrary mode transformation A.

N indistinguishable particles launched into A evolve according to

$$\prod_{j=1}^{N} a^{\dagger}_{\nu_j} |0\rangle \xrightarrow{A} \prod_{j=1}^{N} \left(\sum_{p=1}^{N} A_{p,\nu_j} a^{\dagger}_p \right) |0\rangle . \tag{8.7}$$

Defining $\nu_1 < \nu_2 < \cdots < \nu_N$ and $\mu_1 < \mu_2 < \cdots < \mu_N$ we can consider the transformation A maps the N particles at $\vec{\nu}$ to the output $\vec{\mu}$ in $N!$ indistinguishable possibilities. One of these possibilities maps ν_j to μ_j for all j which we define to not exchange a single pair of particles. The remaining $(N! - 1)$ possibilities map ν_j to σ_{μ_j} for all j for some permutation σ_μ of μ and are defined to exchange particles with the minimum number of pairwise transpositions in clockwise braids that transform ν to σ_μ. Each braid between any two particles makes the combined wave-function acquire a phase ϕ; m clockwise braids that are the minimum number required to map ν to σ_ν therefore cause the global wave-function to acquire phase $m\phi$. From this definition, we obtain an N-particle generalisation of expression (8.5):

$$\left(\langle 0| \prod_{s=1}^{N} a_{\mu_s} \right) \left(\prod_{j=1}^{N} a^{\dagger}_{\nu_j} |0\rangle \right) = \sum_{\sigma_\nu \in S_n} e^{i\tau(\sigma_\nu)\phi} \prod_{j=1}^{N} \delta_{\mu_j, \sigma_{\mu_j}}, \tag{8.8}$$

where $\tau(\sigma_\nu)$ is the minimum number of pairwise transpositions (neighbouring swaps) relating σ_ν to $\vec{\nu}$. Projection onto the output state (8.7) with N-particles at $\vec{\mu}$ uses relations (8.1–8.3) and (8.8) to yield

$$\Gamma^{\phi}_{\vec{\mu}} = \left| \sum_{\sigma_\nu \in S_n} e^{i\tau(\sigma_\nu)\phi} \prod_{j=1}^{n} A_{\mu_j, \sigma_{\nu_j}} \right|^2 \tag{8.9}$$

Varying ϕ allows continuous deformation between Bose–Einstein ($\phi = 0$) and Fermi–Dirac ($\phi = \pi$) quantum interference.

8.3 Simulating Quantum Interference of Two Arbitrary Particles with Entanglement

It is well known that for $\phi = \pi$ and $\phi = 0$, the state

$$\left|\psi^\phi\right\rangle = (|H\rangle_1 |V\rangle_2 + e^{i\phi} |V\rangle_1 |H\rangle_2)/\sqrt{2} \tag{8.10}$$

of two polarised entangled photons incident on a $\eta = 0.5$ reflectivity, non-polarising beam splitter exhibits respectively 'fermion-like' anti-bunching and 'boson-like' bunching. Furthermore, intermediate behaviour [19, 20] has been observed using $|\psi^\phi\rangle$ with varied ϕ incident on a two-mode beamsplitter.

In this section, we show that quantum interference of states of the form $|\psi^\phi\rangle$, unlike number entangled states for example, does generalise to a simulate quantum interference in an arbitrary mode transformation A by sharing the entanglement across two copies of A and observing correlated output. This scheme generalises to simulate N-particle quantum interference using N-partite, N-level entanglement, parameterised by phase ϕ, across N copies of A (Fig. 8.1b). The correlated detection statistics are identical to those arising from N-particle quantum interference with exchange statistics parameter equal to ϕ. This approach could be implemented by any quantum platform that can generate suitable entanglement and evolve the entangled state according to A, and we provide a proof that the generalised state can be generated with a circuit that scales polynomially in N.

8.3.1 Two-Particle Simulation

To physically simulate quantum interference given by the two particle correlation function $\Gamma_{s,t}^\phi$ (Eq. (8.6)), we use two copies of A, denoted A^a and A^b, together with bi-partite entanglement of the form

$$|\psi(\phi)\rangle = (a_j^\dagger b_k^\dagger + e^{i\phi} a_k^\dagger b_j^\dagger) |0\rangle /\sqrt{2} \tag{8.11}$$

where creation operators $a_j^\dagger$ and $b_k^\dagger$ are respectively associated to identical networks A^a and A^b; creation operator sub-indices denote the modes on which the creation operators act. Once $|\psi(\phi)\rangle$ is generated, the class of input entangled particles becomes irrelevant; only one particle is present in each of A^a and A^b, therefore no further multi-particle quantum interference between particles will occur. Evolution of the two particle state through the tensor product of transformations is therefore modelled by

$$|\psi(\phi)\rangle \xrightarrow{A^a \otimes A^b} \sum_{p,q} \left(\frac{A_{p,j}^a A_{q,k}^b + e^{i\phi} A_{p,k}^a A_{q,j}^b}{\sqrt{2}} \right) a_p^\dagger b_q^\dagger |0\rangle \tag{8.12}$$

Fig. 8.1 Simulating quantum interference of quantum particles. **a** Quantum interference of N identical particles with arbitrary statistics launched into a network described by matrix A leads to quantum correlated detection events at the output. **b** Experimental simulation of quantum interference can be directly achieved by sharing N-partite, N-level entanglement across N copies of A, where the choice of simulated particle statistics is controlled by the phase of initial entanglement. Figure reproduced from [1]

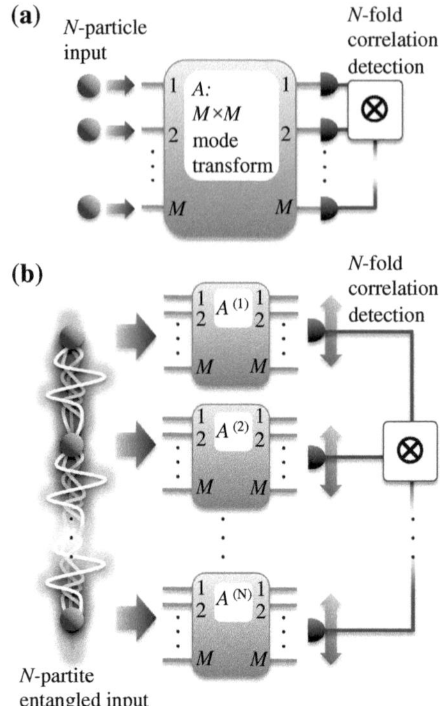

It follows that correlated detection probability across outputs s of A^a and t of A^b simulates Eq. (8.6) according to

$$
P_{s,t}^{\phi} = \frac{1}{2} \Big| \langle 0 | a_s b_t \sum_{p,q} \Big(A_{p,j}^{a} A_{q,k}^{b} + e^{i\phi} A_{p,k}^{a} A_{q,j}^{b} \Big) a_p^{\dagger} b_q^{\dagger} | 0 \rangle \Big|^2
$$

$$
= \frac{1}{2} \Big| A_{s,j}^{a} A_{t,k}^{b} + e^{i\phi} A_{s,k}^{a} A_{t,j}^{b} \Big|^2 \equiv \frac{1}{2} \Gamma_{s,t}^{\phi} \tag{8.13}
$$

where the $1/2$ pre-factor is due to normalisation of the initial entanglement. Note that this treatment agrees with proposals such as [21] that the pattern of correlated detection from entangling the coins of two discrete time quantum walks (a particular unitary process) according to $|\psi^-\rangle$ and $|\psi^+\rangle$ can also be described by Fermi–Dirac and Bose–Einstein statistics respectively.

8.3.2 N-Particle Simulation

Multi-partite entanglement generalises to simulate quantum interference of an arbitrary number of N identical non-interacting particles. Quantum interference of

N indistinguishable particles with exchange statistics ϕ, subjected to mode transformation A at inputs $\vec{\nu} = \{\nu_1, \nu_2, \ldots, \nu_N\}$, are correlated across outputs $\vec{\mu} = \{\mu_1, \mu_2, \ldots, \mu_N\}$ according to

$$\Gamma_{\vec{\mu}}^{\phi} = \left| \sum_{\sigma_\nu \in S_N} e^{i\tau(\sigma_\nu)\phi} \prod_{j=1}^{N} A_{\mu_j, \sigma_{\nu_j}} \right|^2, \tag{8.14}$$

where σ_ν is an element of the permutation group S_N acting on tuple $\vec{\nu}$ and $\tau(\sigma_\nu)$ is the minimum number of pairwise swaps that map σ_ν to $\vec{\nu}$; $\sigma_{\nu j}$ denotes the element j of σ_ν. For $\phi = 0$ and $\phi = \pi$, expression (8.14) is the modulus-square for respectively the permanent and determinant of an $N \times N$ subset of A which is derived for boson and fermion multi-particle quantum walks in Ref. [18] and derived in the appendix for $0 < \phi < \pi$. The multi-particle correlation function (8.14) is simulated by subjecting a generalisation of the entangled N-partite, N-level state

$$\frac{1}{\sqrt{N!}} \sum_{\sigma_\nu \in S_N} e^{i\tau(\sigma_\nu)\phi} \prod_{j=1}^{N} a_{\sigma_{\nu j}}^{(j)\dagger} |0\rangle, \tag{8.15}$$

to into N copies of A such that one particle $a^{(j)\dagger}$ is launched into each copy $A^{(j)}$. N-fold coincidence detection across outputs μ_j of $A^{(j)}$ for $j = 1 \ldots N$ equals $\frac{1}{N!}\Gamma_{\vec{\mu}}^{\phi}$. Note as for the case with two particles, the $1/N!$ pre-factor can be canceled by including all $N!$ permutations of mapping the output tuple $\vec{\mu}$ to the set of copies of the mode transformation $A^{(j)}$, therefore directly simulating quantum interference of N bosons ($\phi = 0$) or N fermions ($\phi = \pi$). All other values of ϕ allow continuous transition between Fermi and Bose quantum interference. In the next section we provide a scalable circuit based on local and controlled operations to generate such states.

8.3.3 Circuit for Generating N-Partite, N-Level Entanglement

A quantum algorithm for anti-symmetrising arbitrary products of n wave functions $|\rho_1, \rho_2, \ldots, \rho_n\rangle_A$ is presented [22] that would use three registers of n, m-digit quantum words or "qu-words" (strings of $n \log_2 m$ qubits). This was then simplified [23] to use two registers of qubits A and B; the first encodes the wavefunctions $|\rho_1, \rho_2, \ldots, \rho_n\rangle$. The second register is mapped to a symmetrized state with $O(n(\ln m)^2)$ local operations followed by $O(n^2 \ln m)$ controlled operations leaving the second register in the state

$$\frac{1}{\sqrt{N!}} \sum_{\sigma \in S_N} |\sigma(1, \ldots, N)\rangle_B \tag{8.16}$$

where $\sigma(1, \ldots, N)$ is a permutation $\sigma \in S_N$ of the N-tuple $\{1, 2, \ldots, N\}$. The state Eq. (8.16) is then antisymmetrised by a sorting algorithm using $O(n \ln n)$ operations, applying the same swaps (via controlled operations) and phase shifts to the first register to antisymmetrize $|\rho_1, \rho_2, \ldots, \rho_n\rangle_A$.

For our purpose, we require only the one register B, since the state we wish to generate is known and of the form

$$|\psi_N(\phi)\rangle = \frac{1}{\sqrt{N!}} \sum_{\sigma_\nu \in S_N} e^{i\tau(\sigma_\nu)\phi} \prod_{j=1}^{N} a_{\sigma_\nu j}^{(j)\dagger} |0\rangle$$

$$= \frac{1}{\sqrt{N!}} \sum_{\sigma_\nu \in S_N} e^{i\tau(\sigma_\nu)\phi} |\sigma_\nu(1, 2, \ldots, N)\rangle \tag{8.17}$$

where $\sigma(1, \ldots, N)$ can be decomposed into $\tau(\sigma)$ adjacent transpositions.

The approach shown in Fig. 8.2 represents each ith qu-word (denoted $B[i]$) in the register as an N-level qudit, for the examples $N = 2, 3$, and generates the state $|\psi_N(\phi)\rangle$ in an iterative process, using $\sum_{i=1}^{N-1} i \sum_{j=1}^{i} j < N^3$ controlled operations. Here each controlled operation is counted as one swap operation of two modes in a given qudit, conditional on an excitation in one of the N levels of the control qudit. Realisation with linear optics for example, could use a heralded KLM CNOT gate [24, 25], each requiring two extra (ancilla) photons, and an interferometric structure of seven beamsplitters.

As with Refs. [22, 23], the N qudits are initialised in superposition using local operations (mode splitters; MS)

$$|1\rangle_{B[N]} \otimes \cdots \otimes |1\rangle_{B[1]} \tag{8.18}$$

$$\xrightarrow{MS} \frac{1}{\sqrt{N!}} \left(\sum_{k=1}^{N} |k\rangle_{B[N]} \right) \otimes \left(\sum_{k=1}^{N-1} |k\rangle_{B[N-1]} \right) \otimes \cdots \otimes |1\rangle_{B[1]}$$

Note that this step puts the state in the correct number of $N!$ terms required for the eventual entangled state. Using linear optics, this would use one photon for each qudit and $\sum_{j=1}^{N}(j - 1) = O(N^2)$ two-mode beam splitters (with appropriate reflectivities).

The next step in generating $|\psi_N(\phi)\rangle$ is to introduce the phase relationship ϕ according to

$$\frac{1}{\sqrt{N!}} \left(\sum_{k=1}^{N} e^{i(k-1)\phi} |k\rangle_{B[N]} \right) \otimes \left(\sum_{k=1}^{N-1} e^{i(k-1)\phi} |k\rangle_{B[N-1]} \right) \otimes \cdots \otimes |1\rangle_{B[1]} \tag{8.19}$$

which uses $\sum_{j=1}^{N}(j - 1) = O(N^2)$ local phase shift operations. A set of $\sum_{i=1}^{N-1} i$ $\sum_{j=1}^{i} j = O(N^4)$ controlled operations are then applied iteratively to generate $|\psi_N(\phi)\rangle$; this process can be thought of as performing a series of controlled operations

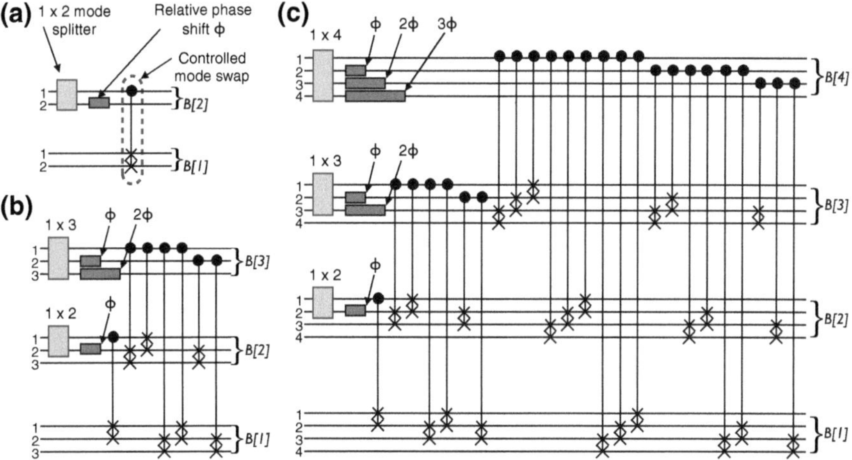

Fig. 8.2 Qudit circuits for generating $|\psi_N(\phi)\rangle$ for **a** $N = 2$ and **b** $N = 3$ and **c** $N = 4$. Qudits are initialised in the state $|1\rangle_{B[N]} \otimes \cdots \otimes |1\rangle_{B[1]}$, where $B[i]$ denotes the ith qudit in the register. The *green squares* represent $1 \times k$ mode splitters that can be decomposed into $k - 1$ two-mode splitters. *Blue rectangles* represent a relative phase shift with labelled phase $j\phi$. Each controlled swap between two modes denoted '$\times$' are conditional on an excitation in the corresponding mode denoted by a '$\bullet$'. Figure reproduced from Ref. [1]

on the intermediate states

$$\frac{1}{\sqrt{q+1}} \sum_{j=1}^{q+1} e^{i(j-1)\phi} |j\rangle_{B[q+1]} \otimes |\psi_q(\phi)\rangle_{B[q,\dots,1]} \tag{8.20}$$

for each step of the iteration $q = 2, \dots, N - 1$. This is verified for $q = 2$ (see Fig. 8.2a and b). Next we assume inductively the process works for some $q = r$ to generate $|\psi_q(\phi)\rangle$. For $q = r + 1$, we begin with the state

$$\frac{1}{\sqrt{r+1}} \sum_{j=1}^{r+1} e^{i(j-1)\phi} |j\rangle_{B[r+1]} \otimes |\psi_r(\phi)\rangle_{B[r,\dots,1]} \tag{8.21}$$

$$= \frac{1}{\sqrt{r+1}} \sum_{j=1}^{r+1} e^{i(j-1)\phi} |j\rangle_{B[r+1]} \otimes \frac{1}{\sqrt{r!}} \sum_{\sigma \in S_r} |\sigma(1, \dots, r)\rangle_{B[r,\dots,1]}$$

A series of controlled qudit shift operations are then applied, such that an excitation of mode j of the $B[r + 1]$ qudit shifts each of the labeled modes $\{j, \dots, r\}$ to $\{j + 1, \dots, r + 1\}$ in the target qudits labelled $\{B[r], \dots B[1]\}$, such that the overall state evolves to

$$\sum_{\sigma \in S_r} e^{i\tau_\nu(\sigma)\phi} \Bigg\{ |1\rangle_{B[r+1]} |\sigma(2, \ldots, r+1)\rangle_{B[r,\ldots,1]} + e^{i\phi} |2\rangle_{B[r+1]}$$

$$|\sigma(1, 3, 4, \ldots, r+1)\rangle_{B[r,\ldots,1]} \cdots + e^{i(j-1)\phi} |j\rangle_{B[r+1]}$$

$$|\sigma(1, 2, 3, \ldots, j-1, j+1, \ldots, r+1)\rangle_{B[r,\ldots,1]}$$

$$+ \cdots + e^{ir\phi} |r+1\rangle_{B[r+1]} |\sigma(1, 2, \ldots, r)\rangle_{B[r,\ldots,1]} \Bigg\} / \sqrt{(r+1)!}$$

$$(8.22)$$

The tuple $\{j, 1, 2, \ldots, j-1, j+1, \ldots, r+1\}$ is $(j-1)$ adjacent transpositions from the tuple $\{1, 2, \ldots, r+1\}$. Since the permutation $\sigma(1, 2, \ldots, j-1, j+1, \ldots, r+1)$ is $\chi(\sigma)$ adjacent transpositions away from the ordered set $\{1, 2, \ldots, j-1, j+1, \ldots, r+1\}$, it follows that the tuple $\{j, \sigma(1, 2, \ldots, j-1, j+1, \ldots, r+1)\}$ is $(j-1+\chi(\sigma))$ adjacent transpositions different from $\{1, 2, \ldots, r+1\}$. Hence the Eq. (8.22) is equivalent to the state

$$|\psi_{r+1}(\phi)\rangle = \frac{1}{\sqrt{(r+1)!}} \sum_{\gamma \in S_r} e^{i\tau_\nu(\gamma)\phi} |\gamma(1, \ldots, r+1)\rangle_{B[r+1,\ldots,1]} \qquad (8.23)$$

as required. Therefore, by induction we have an iterative process that generates $|\psi_N(\phi)\rangle$ for arbitrary size $q = N$, requiring $O(N^2)$ local operations, and $O(N^4)$ controlled swap operations.

8.4 Experimental Realisation Two Particle Quantum Interference Simulation

Here we report experimental simulation of quantum interference of $N = 2$ particles with Fermi–Dirac statistics, Bose–Einstein statistics and fractional statistics for $\phi = \pi/4$, $\phi = \pi/2$ and $\phi = 3\pi/4$. The simulated interference takes place in a continuous time quantum walk unitary (described in Chap. 7) of ten discrete spatial modes and is achieved by sharing post-selected entanglement of two photons shared across two copies of the quantum walk unitary.

We experimentally simulated the correlation function $\Gamma^\phi_{s,t}$ for two arbitrary indistinguishable particles subjected to a coupled oscillator hamiltonian (see Chap. 7) by measuring correlations of post-selected polarisation entanglement of the form $(|H\rangle_j |V\rangle_k + e^{i\phi} |V\rangle_j |H\rangle_k)/\sqrt{2}$ shared across two process copies A^a and A^b. These copies are given by the TE and TM modes [26] of birefringent silicon oxynitride (SiON) waveguide array (Fig. 8.3b) that realize two independent continuous time quantum walks [27]. The waveguides and butt-coupled polarisation maintaining fibres (PMF) are birefringent to ensure negligible cross-talk between the horizontal and vertical polarisation. Detection statistics of the output photons in the $\{|H\rangle, |V\rangle\}$ basis were analysed using polarisation beam splitters (PBS) and pairs

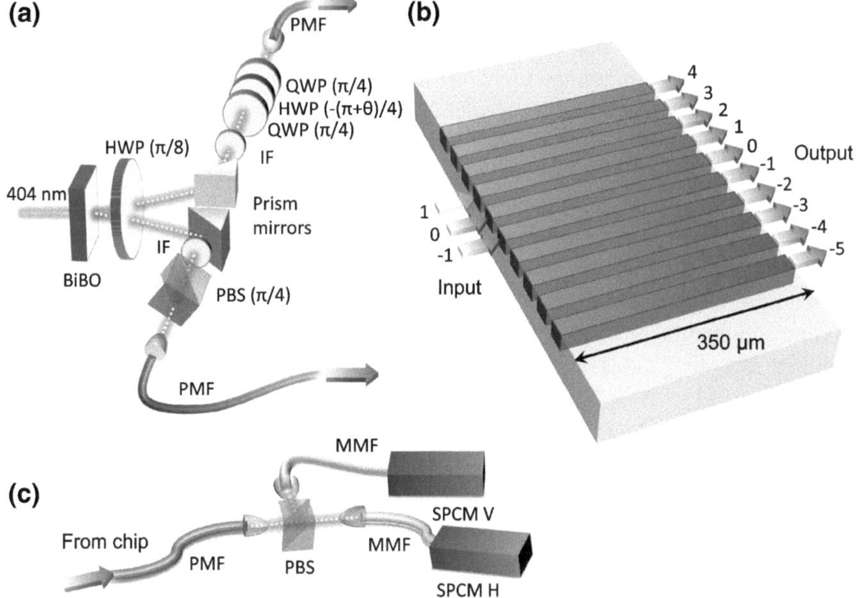

Fig. 8.3 Experimental setup **a** The parametric down-conversion based photon pair source with quarter-waveplates (QWP), half-waveplates (HWP) and a polarisation beamsplitter cube (PBS) used as a polariser, to control the initial state before input into the chip. Down conversion in the nonlinear bismuth borate (BiBO) yields degenerate photon pairs (808 nm, filtered with 2 nm full width, half maximum interference filters, IF) collected into polarisation maintaining fibres (PMF). See appendix for further detail. **b** Two copies of the quantum walk unitary are realised by accessing the TE and TM modes of the polarisation preserving waveguide. Three input modes of the array are used in the experiment $(-1, 0, 1)$ accessed via waveguide bends that fan to a pitch of 250 μm and butt-coupled to an array of PMF. Output is accessed via a fan of waveguide from the coupling region to a pitch of 125 μm. **c** The detection scheme consists of five PBS and single photon counting module (SPCM) pairs. PMF coupled to the chip launch the output photons onto each PBS which separates TE and TM modes guided in the waveguide. The two outputs of each PBS are collected into multi-mode fibre (MMF) coupled to SPCM for correlated detection via counting electronics. Figure reproduced from Ref. [1]

of multi-mode fibre coupled avalanche photo-diode single photon counting modules (SPCM, PerkinElmer) (Fig. 8.3c). From only coincidental events occurring from detecting one $|H\rangle$ photon and one $|V\rangle$ photon at any combination of two outputs of the chip, the input state $(|H\rangle_j |V\rangle_k + e^{i(\phi+\epsilon)} |V\rangle_j |H\rangle_k)/\sqrt{2}$ was post-selected with a measurable phase offset ϵ (from differing lengths of the two birefringent PMF and the input waveguides) from the initial state $(|H\rangle_j + |V\rangle_j)(|H\rangle_k + e^{-i\phi} |V\rangle_k)/2$, prepared with quarter- and half-waveplates. The phase ϵ is compensated modulo 2π using the $Z(\theta)$ polarisation phase rotation that also sets ϕ and is comprised of two quarter waveplates (QWP) and one HWP as shown in Fig. 8.3a. Note that since both terms in the post selected state contain one H photon and one V photon, birefringence of the $\sim$2 m PMF does not cause decoherence of the desired state.

The photon pairs were generated in a 2 mm thick, nonlinear bismuth borate BiB_3O_6 (BiBO) crystal phase matched for non-colinear type-I spontaneous parametric downconversion, pumped by a 60 mW 404 nm continuous wave diode laser. The photons were spectrally filtered using 2 nm full width at half maximum interference filters centred on 808 nm (Semrock) and collected with aspheric lens focusing on polarisation maintaining fibres. A typical two-photon count rate of 50 kHz was measured directly from the source with two $\sim$60 % efficient SPCM. For a coherent input state, arrival time of the two photons at the input of the waveguide array is made identical via characterisation with a Hong–Ou–Mandel experiment [28].

Two sets of correlation matrices were measured for the two pairs of input with maximum quantum interference overlap in the waveguide array [11]: Immediate neighbouring waveguides in the centre of the array $(j, k) = (-1, 0)$ and waveguides $(j, k) = (-1, 1)$ with vacuum state in waveguide 0. The resulting correlation matrices with $P_{s,t}^{\phi}$ entry given by coincidental detection of respectively horizontal and vertical polarised photons at outputs s and t for input $(-1, 0)$ is given in Fig. 8.4. Five cases are displayed for respectively $\phi = 0, \pi/4, \pi/2, 3\pi/4, \pi$ for both experimental measurement (left: a, c, e, g, i) and the ideal distribution given by $\Gamma_{s,t}^{\phi}$ in Eq. (8.6) (right: b, d, f, h, j). We observe the expected bunching behaviour of bosons for $\phi = 0$ which is contrasted with the observation of the Pauli-exclusion in the diagonal elements of the matrix for $\phi = \pi$. We note that a form of exclusion on the diagonal entries $P_{j,j}^{\phi}$ would also occur for experiments with hard core bosons, but the remaining correlation matrix would still be dependent upon ϕ. The spectrum of phase $0 < \phi < \pi$ exhibits a variable Pauli-exclusion and a continuous transition between simulation of Bose–Einstein and Fermi–Dirac quantum interference.

The correlation matrices $P_{s,t}^{\phi}$ for the input $(-1, 1)$ is given in Fig. 8.5 with experimental measurements displayed (left: a, c, e, g, i) together with the ideal distributions for a range of particle statistics (right: b, d, f, h, j) ranging from Boson to Fermion for the phases $\phi = 0, \pi/4, \pi/2, 3\pi/4, \pi$. Unlike the -1 and 0 input for $\phi = 0$ bunching behaviour was not as prevalent over other populations of the correlation matrix: we observed the expected correlation behaviour [29] for this input. As we varied ϕ, we observed transition between simulation of Bose–Einstein and Fermi–Dirac quantum interference.

Agreement between an ideal model that assumes a perfect device and experimental measurement is quantified in Table 8.1 using the similarity (S) between the ideal correlation matrix Γ^{ϕ} and the measured distribution of detection statistics P^{ϕ}, defined [11] by $S = (\sum_{i,j} \sqrt{\Gamma_{i,j}^{\phi} P_{i,j}^{\phi}})^2 / \sum_{i,j} \Gamma_{i,j}^{\phi} \sum_{i,j} P_{i,j}^{\phi}$; a generalisation of the average (classical) fidelity between two probability distributions that ranges from 0 % (anti-correlated distributions) to 100 % for complete agreement. The temporal modes of the two inputs were also made distinguishable by relative delay of the photon pairs to outside of the coherence time. This allowed physical simulation of the classical correlations $\Gamma^C = |A_{s,j} A_{t,k}|^2 + |A_{s,k} A_{t,s}|^2$ that are independent of particle statistics. For these measurements, we found agreement with the ideal model of $S = 97.4 \pm 0.7$ % and $S = 97.5 \pm 0.6$ % for inputs $(-1, 0)$ and $(-1, 1)$

Fig. 8.4 Experimental simulation of particle statistics correlation functions for input $(-1, 0)$. Experimental correlation matrices (*left* **a, c, e, g, i**) for simulating quantum interference of non-interacting particle pairs with five different particle statistics. Ideal correlation matrices for this experiment are plotted (*right* **b, d, f, h, j**) using Eq. (8.12). The data ranges from corresponding to Bose–Einstein statistics ($\phi = 0$, (**a, b**)) through fractional statistics ($\phi = \pi/4$, (**c, d**); $\phi = \pi/2$, (**e, f**); $\phi = 3\pi/4$, (**g, h**)) to Fermi–Dirac statistics ($\phi = \pi$, (**i, j**)). **k** The correlation matrices are displayed as normalised probability distributions with *white* for probability $P^{\phi}_{i,j} = 0$ and *black* for $P^{\phi}_{i,j} = 1$. *Solid coloured squares* represent measured correlations. Coincidence rates are corrected for relative detection efficiency. Figure reproduced from Ref. [1]

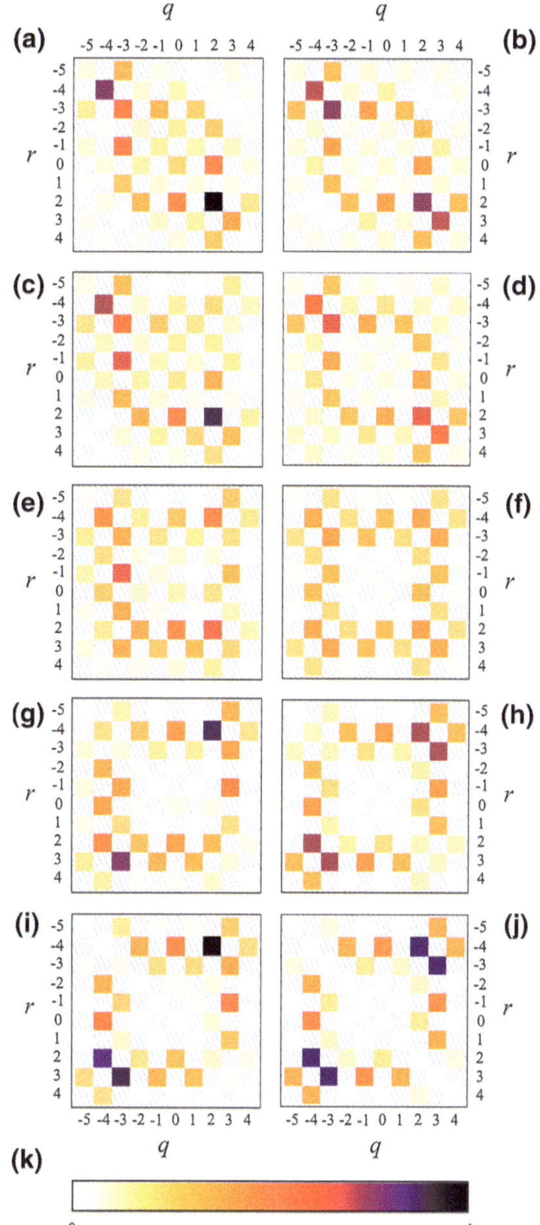

respectively. Our results indicate A^a and A^b are near identical while keeping crosstalk between horizontal and vertical polarisation to a minimum. While non perfect similarity ($S = 100\,\%$) in Table 8.1 is attributed to non-perfect entanglement generation,

Fig. 8.5 Experimental simulation of particle statistics correlation functions for input $(-1, 1)$ with data laid out as for Fig. 8.4. Figure reproduced from Ref. [1]

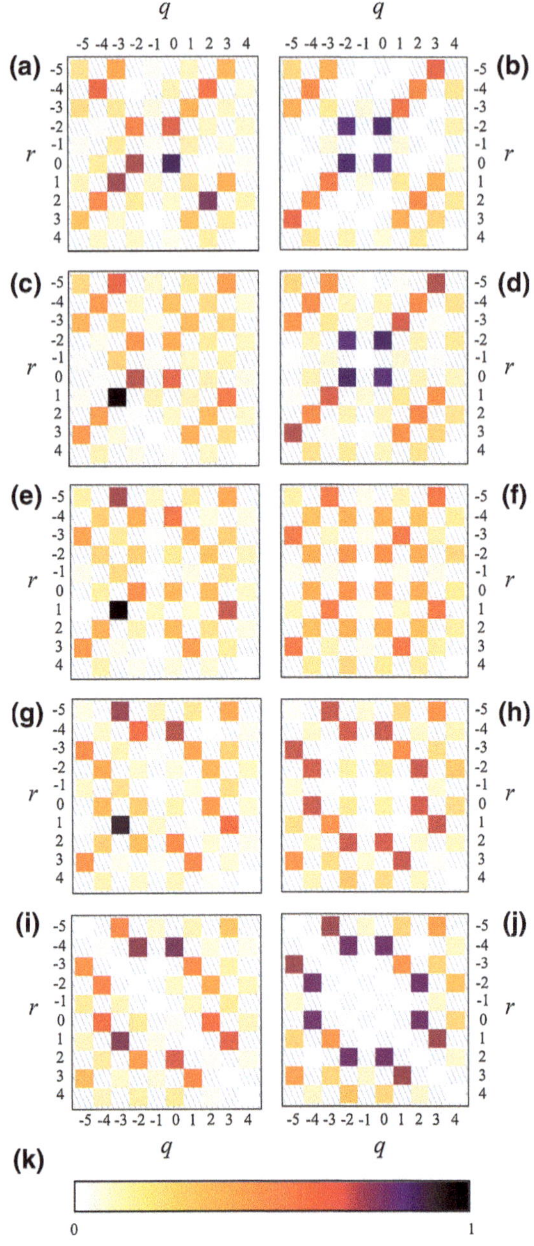

this level of agreement between the ideal case and measurement indicates accurate quantum simulation of arbitrary quantum interference in a uniform coupled oscillator hamiltonian.

Table 8.1 Similarity S for each experimental simulation of particle quantum interference with parameter ϕ, for input $(-1, 0)$ given in Fig. 8.4 and input $(-1, 1)$ given in Fig. 8.5

Phase, ϕ	S (%) for input $(-1, 0)$	S (%) for input $(-1, 1)$
0	95.8 ± 0.6	90.9 ± 0.5
$\pi/4$	95.2 ± 0.7	93.6 ± 0.6
$\pi/2$	94.1 ± 0.7	93.5 ± 0.6
$3\pi/4$	94.4 ± 0.7	94.1 ± 0.6
π	91.6 ± 0.7	92.7 ± 0.6

8.5 Discussion

We have shown that the algebra for certain entanglement evolving according to some operation A directly maps onto the quantum interference of indistinguishable particles in one copy of A with a direct correspondence between a single phase parameter of the entangled input and to the exchange statistics of quantum particles. This can be considered as an analogue quantum simulation of arbitrary quantum interference.

We have experimentally demonstrated this idea using post-selected polarisation entangled photons, undergoing evolution in a quantum walk unitary, which can also be interpreted as two entangled continuous time quantum walks.[1] Furthermore, the fermion simulation directly maps to the dynamics of two excitations in a spin$-\frac{1}{2}$ chain consisting of ten sites: Using the Jordan Wigner transformation, excitations correspond to fermion creation operators and the vacuum to the ground state of each spin [31]. Our simulation of arbitrary fractional exchange statistics opens the way to a range of anyon quantum interference phenomena and compliments recent work using optics to study fixed fractional phase in Kitaev's Hamiltonian [32, 33].

We have theoretically shown that this approach generalises to simulating an arbitrary number of particles, using qudit entanglement. Furthermore, we have shown that the required state can in principle be generated with a circuit of controlled operations that scales polynomially in N. The ability to simulate arbitrary particle statistics also has immediate applications, using photons guided in inherently stable three-dimensional waveguide architecture [34] for example for quantum chemistry [35] and in combinatorics where exchange statistics have been shown to alter graph topology [36].

The method we report is not reliant on using polarisation and can be applied to any quantum process realisable with linear optics [37]. The entangled state, either generated deterministically or via post-selection, could be encoded in time-bin [38] or interferometrically stable path with reconfigurable waveguide circuitry [39]. Entanglement can be applied as described here to simulate large scale quantum interference of arbitrary quantum particles with quantum platforms outside of photonics.

---

[1] A similar experiment using two polarisation entangled photons in a discrete time quantum walk, realized with a network of integrated directional couplers, was recently reported on the arXiv [30].

References

1. J.C.F. Matthews, K. Poulios, J.D.A. Meinecke, A. Politi, A. Peruzzo, N. Ismail, K. Wörhoff, M.G. Thompson, J.L. O'Brien, Simulating quantum statistics with entangled photons: a continuous transition from bosons to fermions. arxiv:1106.1166 (2011).
2. Y.L. Lim, A. Beige, Generalized Hong-Ou-Mandel experiments with bosons and fermions. New J. Phys. **7**, 155 (2005)
3. J.D. Franson, B.C. Jacobs, T.B. Pittman, Quantum computing using single photons and the zeno effect. Phys. Rev. A **70**(6), 062302 (2004)
4. K. Kim, M.-S. Chang, S. Korenblit, R. Islam, E.E. Edwards, J.K. Freericks, G.-D. Lin, L.-M. Duan, C. Monroe, Quantum simulation of frustrated ising spins with trapped ions. Nature **465**, 590–593 (2010)
5. J.T. Barreiro, M. Müller, P. Schindler, D. Nigg, T. Monz, M. Chwalla, M. Hennrich, C.F. Roos, P. Zoller, R. Blatt, An open-system quantum simulator with trapped ions. Nature **470**, 486–491 (2011)
6. F.E. Camino, W. Zhou, V.J. Goldman, Realization of a laughlin quasiparticle interferometer: observation of fractional statistics. Phys. Rev. B **72**, 075342 (2005)
7. D. Loss, D.P. DiVincenzo, Quantum computation with quantum dots. Phys. Rev. A **57**, 120–126 (1998)
8. T. Byrnes, N.Y. Kim, K. Kusudo, Y. Yamamoto, Quantum simulation of Fermi-Hubbard models in semiconductor quantum-dot arrays. Phys. Rev. B **78**, 075320 (2008)
9. J.F. Sherson, C. Weitenberg, M. Endres, M. Cheneau, I. Bloch, S. Kuhr, Single-atom-resolved flourecence imaging of an atomoc mott insulator. Nature **467**, 68–72 (2010)
10. S. Aaronson, A. Arkhipov, The computational complexity of linear optics, in Proceedings of ACM STOC 2011. arXiv:1011.3245 [quant-ph] (2010) (to appear).
11. A. Peruzzo, M. Lobino, J.C.F. Matthews, N. Matsuda, A. Politi, K. Poulios, X.-Q. Zhou, Y. Lahini, N. Ismail, K. Wörhoff, Y. Bromberg, Y. Silberberg, M.G. Thompson, J.L. O'Brien, Quantum walks of correlated photons. Science **329**, 1500–1503 (2010)
12. R. Loudon, Fermion and boson beam-splitter statistics. Phys. Rev. A **58**, 4904 (1998)
13. R.C. Liu, B. Odom, Y. Yamamoto, S. Tarucha, Quantum interference in electron collision. Nature **391**, 263–265 (1998)
14. F. Wilczek, Magnetic flux, angular momentum and statistics. Phys. Rev. Lett. **48**, 1144 (1982)
15. F. Wilczek, *Fractional Statistics and Anyon Superconductivity* (World Scientific, Singapore, 1990)
16. R.B. Laughlin, Anomalous quantum hall effect: an incompressible quanutm fluid with fractionaly charged excitation. Phys. Rev. Lett. **50**, 1395–1398 (1983)
17. K. Mattle, M. Michler, H. Weinfurter, A. Zeilinger, M. Zukowski, Non-classical statistics at multiport beam splitter. Appl. Phys. B Lasers Opt. 60(2–3), Suppl. S, S111 (1995).
18. K. Mayer, M.C. Tichy, F. Mintrt, T. Konrad, A. Buchleitner, Counting statistics of many-particle quantum walks. Phys. Rev. A **83**, 061207 (2011)
19. M. Michler, K. Mattle, H. Weinfurter, A. Zeilinger, Interferometric bell-state analysis. Phys. Rev. A **53**, R1209–R1212 (1996)
20. Q. Wang, Y.-S. Zhang, Y.-F. Huang, G.-C. Guo, Simulating the fourth-order interference phenomenon of anyons with photon pairs. Eur. Phys. J. D. **42**, 179–182 (2007)
21. Y. Omar, N. Paunković, L. Sheridan, S. Bose, Quantum walk on a line with two entangled particles. Phys. Rev. A **74**, 042304 (2006)
22. D.S. Abrams, S. Lloyd, Simulation of many-body fermi systems on a universal quantum computer. Phys. Rev. Lett. **79**, 2586–2589 (1997)
23. N.J. Ward, I. Kassal, A. Aspuru-Guzik, Preparation of many-body states for quantum simulation. J. Chem. Phys. **130**, 194105 (2009)
24. T.C. Ralph, A.G. White, W.J. Munro, G.J. Milburn, Simple scheme for efficient linear optics quantum gates. Phys. Rev. A **65**, 012314 (2001)

25. R. Oxalmoto, J.L. O'Brien, H.F. Hofmann, S. Takeuchi, Realization of a knill-laflamme-milburn controlled-not photonic quantum circuit combing effective optical nonlinearities. PNAS **108**(25), 10067–10071 (2011)
26. L. Sansoni, F. Sciarrino, G. Vallone, P. Mataloni, A. Crespi, R. Ramponi, R. Osellame, Polarization entangled state measurement on a chip. Phys. Rev. Lett. **105**, 200503 (2010)
27. H.B. Perets, Y. Lahini, F. Pozzi, M. Sorel, R. Morandotti, Y. Silberberg, Realization of quantum walks with negligible decoherence in waveguide lattices. Phys. Rev. Lett. **100**(17), 170506 (2008)
28. C.K. Hong, Z.Y. Ou, L. Mandel, Measurement of subpicosecond time intervals between two photons by interference. Phys. Rev. Lett. **59**(18), 2044–2046 (1987)
29. Y. Bromberg, Y. Lahini, R. Morandotti, Y. Silberberg, Quantum and classical correlations in waveguide lattices. Phys. Rev. Lett. **102**(25), 253904 (2009)
30. L. Sansoni, F. Sciarrino, G. Vallone, P. Mataloni, A. Crespi, R. Ramponi, R. Osellame, Two-particle bosoinic-fermionic quantum walk via integrated photonics. Phys. Rev. Lett. **108**, 010502 (2012)
31. A.P. Hines, P.C.E. Stamp, Quantum walks, quantum gates, and quantum computers. Phys. Rev. A **75**, 062321 (2007)
32. C.-Y. Lu, W.-B. Gao, O. Guhne, X.-Q. Zhou, Z.B. Chen, J.-W. Pan, Demonstrating anyonic fractional statistics with a six-qubit quantum simulator. Phys. Rev. Lett. **102**, 030502 (2009)
33. J.K. Pachos, W. Wieczorek, C. Schmid, N. Kiesel, R. Pohlner, H. Weinfurter, Revealing anyonic features in a toric code quantum simulation. New J. Phys. **11**, 083010 (2009)
34. E. Keil, A. Szameit, F. Dreisow, M. Heinrich, S. Nolte, A. Tünnermann, Photon correlations in two-dimensional waveguide arrays and their classical estimate. Phys. Rev. A **81**, 023834 (2010)
35. I. Kassal, S.P. Jordan, P.J. Love, M. Mohseni, A. Aspuru-Guzik, Polynomial-time quantum algorithm for the simulation of chemical dynamics. Proc. Natl. Acad. Scie. USA **105**, 18681–18686 (2008)
36. J.M. Harrison, J.P. Keating, J.M. Robbins, Quantum statistics on graphs. Proc. R. Soc. A **467**, 212–233 (2011)
37. M. Reck, A. Zeilinger, H.J. Bernstein, P. Bertani, Experimental realization of an discrete unitary operator. Phys. Rev. Lett. **73**(1), 58–61 (1994)
38. R.T. Thew, A. Acin, H. Zbinden, N. Gisin, Bell-type test of energy-time entangled qutrits. Phys. Rev. Lett. **93**, 010503 (2004)
39. J.C.F. Matthews, A. Politi, A. Stefanov, J.L. O'Brien, Manipulation of multiphoton entanglement in waveguide quantum circuits. Nat. Photonics **3**, 346–350 (2009)

Chapter 9
Conclusions and Outlook

In this thesis, I have presented the theoretical model, the setup details and results of a range of quantum photonics experiments in integrated waveguide optics that I have performed with collaborators during my PhD studies. Here the key results of the entire thesis are summarised. This is followed with an outlook to future research in the same area.

9.1 Key Results

At the end of each chapter, I have included relevant discussion that puts the results of the chapter into context with respect to the field at large. The key results of the thesis are collected together here.

1. **Quantum interference in directly laser written waveguides**. In Chap. 3, we have shown that the rapid prototyping direct laser writing technique is capable of fabricating waveguide circuits suitable for quantum optical experiments. By observing high quantum interference visibility in the Hong–Ou–Mandel experiment using a range of directional couplers ($V_{max} = 96\,\%$), we have confirmed that all of the properties of guided single photons remains virtually unchanged. Furthermore, by testing for quantum interference in a range of reflectivity couplers we show that the technique is largely repeatable. We further perform a three photon generalisation of the Hong–Ou–Mandel effect.

2. **Demonstration of a compiled version of Shor's algorithm to factorize 15**. We used multiple directional couplers in a lithographically fabricated architecture (silica-on-silicon) to realise a simple quantum logic circuit that uses two linear optical two-qubit logic gates. This demonstration served as a benchmark of the complexity of quantum circuits achievable at the time with waveguide directional couplers.

J. C. F. Matthews, *Multi-Photon Quantum Information Science and Technology in Integrated Optics*, Springer Theses, DOI: 10.1007/978-3-642-32870-1_9,
© Springer-Verlag Berlin Heidelberg 2013

3. **Quantum interference in reconfigurable waveguide circuitry**. We demonstrated a building block for realising fully reconfigurable linear optical circuits by performing multiple Hong–Ou–Mandel experiments in a waveguide Mach Zehnder Interferometer, used as a reconfigurable beam splitter.

4. **Manipulation of entangled states of light using a reconfigurable waveguide interferometer**. We demonstrated the capacity for integrated optics to manipulate entangled states of light suitable for quantum metrology, by observing super-resolved interference fringes of post-selected two- and four-photon NOON states with visibility sufficient to beat the standard quantum limit.

5. **Realisation of a scheme** [1] **for heralding photon number-path entanglement**. We tested the projective measurement of two auxiliary photons after quantum interference of four- and six-photons in a waveguide circuit to herald the generation of respectively two- and four-photon NOON states. The circuit included a variable phase shift (necessary for the six-photon experiment) which we used to verify the heralded generation of a four photon, maximally entangled state that is predicted to be resilient to (balanced) loss, compared to NOON states.

6. **Two photon quantum walks**. Starting from the model of correlated two photon continuous time quantum walks [2], we proved theoretically that two photon quantum interference in a planar waveguide array of size N simulates the continuous time quantum walk of one particle moving around a graph of size N^2; this generalises to performing a quantum walk of n particles to simulate a quantum walk on a n dimensional graph of size $O(N^n/n!)$. We also reported experimental results of a two-photon continuous time quantum walk in a waveguide array.

7. **Simulation of quantum interference of particles with arbitrary statistics using entanglement**. We reported a rigorous theory that the behaviour of certain entanglement exactly maps onto the behaviour of two particles undergoing quantum interference in an arbitrary mode transformation. This is directly controlled by a single phase parameter that maps onto the exchange symmetry of the simulated particles. We generalised this idea to simulate an arbitrary number of particles using entangled qudits and we provide a circuit for generating the required entanglement that scales polynomially in the number of particles generated.

8. **Experimental simulation of quantum interference of bosons, fermions and fractional quasiparticles in the intermediate regime**. Following our derived theory (above), we use polarisation entangled photons launched into a waveguide array to realise two entangled quantum walks in the TE and TM modes of the array, whose correlated behaviour exactly maps onto the correlated quantum interference quantum walks of two identical bosons, fermions. We control the phase of our entanglement to realise behaviour intermediate between these two classes.

9.2 Outlook

Our initial quantum interference results in directly laser written waveguide circuits [3], open the way to realising more complex quantum optical devices. Rapid prototyping of integrated quantum photonic circuits can in principle be achieved with ultrafast laser system as a viable alternative to using state-of-the-art semiconductor processing facilities. In particular, directly laser written waveguide circuits are amenable to integration with more complex laser machined components, including micro-fluidic channels [4]: this may provide accurate refractive index measurement of biological samples using non-classical states of light for example. One of the key features of directly written waveguides is the fabrication of three dimensional structures (for example [5]) and has led to the recent result of a two photon quantum walk in a three-dimensional ring structure [6]. Together, with the ability to simulate arbitrary quantum particle exchange statistics [7] (Chap. 8), this technology could for example be applied to studying multi-particle quantum walks with Fermi–Dirac statistics on two-dimensional graphs [8] and for analogue quantum simulation of idealised solid state systems such spin chains [9].

The results of Chaps. 4 and 5 of manipulating quantum information on chip [10] and for performing simple quantum logic [11] indicate the suitability for realising large scale reconfigurable circuits based on nested interferometers and variable phase shifts. Realising such circuits with even a small number of phase shifts will allow for example state and process tomography of path encoded circuitry, directly on chip. It should be noted that cross talk between thermal phase shifts is likely to be an effect in more complex circuits, and by the very fact that the phase shift takes place via heating, simultaneous application of multiple heaters may likely cause missalignment with input and output optical fibres. The telecommunications industry has, however, demonstrated 32×32 silica based optical switches for example, in a nested interferometer network consisting of $O(10^3)$ thermal phase shifters [12]; this indicates large scale quantum circuits can be realised with thermo optics. With regards to speed, the state of the art in thermal shifting is $O(10^2)\,$ns [13]. This is certainly suitable for reconfigurable quantum circuits [14] and for multiplexed down conversion schemes [15, 16]. Schemes requiring higher switching rates are likely to use other material systems. Lithium niobate, for example, exhibits an electro-optic property that allows sub-nanosecond switching rates. This is more suited to 'feed forward' schemes that require fast optical switching (conditional on certain detection outcomes) such as cluster-state quantum computation schemes currently implemented in bulk optics [17].

Of course, the general quantum behaviour of photons demonstrated in Chaps. 3–5 can be (and have been) observed in bulk optical circuits. One challenge is to now realise quantum optical circuits that, through harnessing the inherent stability and complexity achievable with integrated quantum circuits, will allow quantum optical experiments to be performed that cannot be achieved by other means. This ranges from building integrated quantum circuits constructed from phase shifters and directional couplers in large nested interferometer structures, to exploring novel photonic components that do not have a simple bulk optical analogue. One such architecture is the evanescently coupled waveguide array, reported in Chaps. 7 and 8 to perform ordered quantum walk operations. Through their inherent stability, integrated optics allow straight forward methods for path encoded higher dimensional systems—the two photon quantum walk experiment of Chap. 7 is equivalent to unitary rotation of an 231-level qudit (see Fig. 7.4b). Controlling the individual parameters of the waveguide arrays will allow a multitude of quantum photonic behaviour to be observed.

Waveguide arrays have allowed us to test with a complex unitary operation the idea that arbitrary quantum particle statistics can be simulated with entanglement for an arbitrary mode transformation. Our experimental demonstration uses polarisation entanglement, which according to our scheme is limited to simulating only two particles. However, time-bin or path entangled states of multiple qudits will allow quantum simulation of more particles.

In this thesis, we have used integrated optics purely for realising circuitry, which have provided encouraging results that verify integrated optics as a suitable platform for performing photonic quantum information circuits. The true potential, however, of integration will only be reached when it is possible to integrate components such as single photon sources, single photon detectors together with waveguide circuitry. Multiplexed downconversion sources [15] using for example periodically poled lithium niobate [18] may provide one viable route to generating high photon number sources: such a scheme will benefit from integration with avalanche photo diode detectors [19] or superconducting detectors [20]. Integration with photon sources and detectors on the whole are immediate goals for the field and will further increase practicality and physical scalability of quantum photonic demonstrations for quantum technology and quantum information science.

References

1. H. Lee, P. Kok, N.J. Cerf, J.P. Dowling, Linear optics and projective measurements alone suffice to create large-photon-number path entanglement. Phys. Rev. A **65**, 030101 (2002)
2. Y. Bromberg, Y. Lahini, R. Morandotti, Y. Silberberg, Quantum and classical correlations in waveguide lattices. Phys. Rev. Lett. **102**(25), 253904 (2009)
3. G.D. Marshall, A. Politi, J.C.F. Matthews, P. Dekker, M. Ams, M.J. Withford, J.L. O'Brien, Laser written waveguide photonic quantum circuits. Opt. Express **17**(15), 12546–12554 (2009)
4. K.C. Vishnubhatla, N. Bellini, R. Ramponi, G. Cerullo, R. Osellame, Shape control of microchannels fabricated in fused silica by femtosecond laser irradiation and chemical etching. Opt. Express **17**, 8685 (2009)
5. E. Keil, A. Szameit, F. Dreisow, M. Heinrich, S. Nolte, A. Tünnermann, Photon correlations in two-dimensional waveguide arrays and their classical estimate. Phys. Rev. A **81**, 023834 (2010)
6. J.O. Owens, M.A. Broome, D.N. Biggerstaff, M.E. Coggin, A. Fedrizzi, T. Linjordet, M. Ams, G.D. Marshall, J. Twamley, M.J. Withford, A.G. White, Two-photon quantum walks in an elliptical direct-write waveguide array. New J. Phys. **13**, 075003 (2011)

7. J.C.F. Matthews, K. Poulios, J.D.A. Meinecke, A. Politi, A. Peruzzo, N. Ismail, K.Wörhoff, M.G. Thompson, J.L. O'Brien, Simulating quantum statistics with entangled photons: a continuous transition from bosons to fermions. arxiv:1106.1166 (2011)
8. J.M. Harrison, J.P. Keating, J.M. Robbins, Quantum statistics on graphs. Proc. R. Soc. A **467**, 212–233 (2011)
9. S. Bose, Quanutm communication through an unmodulated spin chain. Phys. Rev. Lett. **91**, 207901 (2003)
10. J.C.F. Matthews, A. Politi, A. Stefanov, J.L. O'Brien, Manipulation of multiphoton entanglement in waveguide quantum circuits. Nat. Photonics **3**, 346–350 (2009)
11. A. Peruzzo, M. Lobino, J.C.F. Matthews, N. Matsuda, A. Politi, K. Poulios, X.-Q. Zhou, Y. Lahini, N. Ismail, K. Wörhoff, Y. Bromberg, Y. Silberberg, M.G. Thompson, J.L. O'Brien, Quantum walks of correlated photons. Science **329**, 1500–1503 (2010)
12. S. Sohma, T. Watanabe, N. Ooba, M. Itoh, T. Shibata, H. Takahashi, Silica-based plc type 32×32 optical matrix switch. *European Conference on Optical Communication*, 2006, pp. 1–2
13. G. Coppola, L. Sirleto, I. Rendina, M. Iodice, Advance in thermo-optical switches: principles, materials, design, and device structure. Opt. Eng. **50**, 071112 (2011)
14. M. Reck, A. Zeilinger, H.J. Bernstein, P. Bertani, Experimental realization of an discrete unitary operator. Phys. Rev. Lett. **73**(1), 58–61 (1994)
15. A.L. Migdall, D. Branning, S. Castelletto, Tailoring single-photon and multiphoton probabilities of a single-photon on-demand source. Phys. Rev. A **66**(5), 053805 (2002)
16. X.-S. Ma, S. Zotter, J. Kofler, T. Jennewein, A. Zeilinger, Experimental genreation of single photons via active multiplexing. Phys. Rev. A **83**, 043814 (2011)
17. R. Prevedel, P. Walther, F. Tiefenbacher, P. Bohi, R. Kaltenbaek, T. Jennewein, A. Zeilinger, High-speed linear optics quantum computing using active feed-forward. Nature **445**(7123), 65–69 (2007)
18. S. Tanzilli, W. Tittel, H. De Riedmatten, H. Zbinden, P. Baldi, M. De Micheli, D. Ostrowsky, N. Gisin, PPLN waveguide for quantum communication. Eur. Phys. J. D **18**(2), 155–160 (2002)
19. K. Tsujino, M. Akiba, M. Sasaki, Ultralow-noise readout circuit with an avalanche photodiode: toward a photon-number-resolving detector. Appl. Opt. **46**(7), 1009–1014 (2007)
20. A. Divochiy, F. Marsili, D. Bitauld, A. Gaggero, R. Leoni, F. Mattioli, A. Korneev, V. Seleznev, N. Kaurova, O. Minaeva, G. Gol'tsman, K.G. Lagoudakis, M. Benkhaoul, F. Levy, A. Fiore, Superconducting nanowire photon-number-resolving detector at telecommunication wavelengths. Nat. Photonics **2**(5), 302–306 (2008)